Lecture Notes in Artificial Intelligence 804

Subseries of Lecture Notes in Computer Science
Edited by J. G. Carbonell and J. Siekmann

Lecture Notes in Computer Science
Edited by G. Goos and J. Hartmanis

Daniel Hernández

Qualitative Representation of Spatial Knowledge

Springer-Verlag

Berlin Heidelberg New York
London Paris Tokyo
Hong Kong Barcelona
Budapest

Series Editors

Jaime G. Carbonell
School of Computer Science, Carnegie Mellon University
Schenley Park, Pittsburgh, PA 15213-3890, USA

Jörg Siekmann
University of Saarland
German Research Center for Artificial Intelligence (DFKI)
Stuhlsatzenhausweg 3, D-66123 Saarbrücken, Germany

Author

Daniel Hernández
Fakultät für Informatik, Technische Universität München
Arcisstraße 21, D-80290 München, Germany

CR Subject Classification (1991): I.2.4, I.2.0, I.2.1, I.2.3, J.6

ISBN 3-540-58058-1 Springer-Verlag Berlin Heidelberg New York
ISBN 0-387-58058-1 Springer-Verlag New York Berlin Heidelberg

CIP data applied for

© Springer-Verlag Berlin Heidelberg 1994
Printed in Germany

Typesetting: Camera ready by author
SPIN: 10131112   45/3140-543210 - Printed on acid-free paper

Preface

Cognitive spatial concepts are *qualitative* in nature, i.e., they are based not so much on exact quantities but on comparisons between perceived magnitudes. We develop a qualitative model for the representation of spatial knowledge (in particular, of positional information about 2-dimensional projections) that is based only on locative relations between the objects involved, does not need a global scale, and takes different kinds of reference frames into consideration. The resulting model is both cognitively plausible and computationally efficient.

The core of the book is the study of qualitative inference methods. We extend constraint reasoning mechanisms to take the rich structure of physical space into consideration, and describe methods to transform among reference frames and for the composition of relations. We also sketch extensions of the qualitative approach to represent positional information in 3-D scenes, as well as other spatial concepts such as size, shape, and distance.

Qualitative representations make only as many distinctions as necessary to identify objects, events, situations, etc. in a given context (recognition task) as opposed to those needed to fully reconstruct a situation (reconstruction task). Thus, they can be used in many application areas from everyday life in which spatial knowledge plays a role, particularly in those that are characterized by uncertain and incomplete knowledge, such as computer aided systems for architectural design, geographical information systems, but also robot control or natural language information systems to give directions.

This book is a revised version of a dissertation submitted to the *Institut für Informatik* (Department of Computer Science) of the *Technische Universität München* (Munich, Germany) in December of 1992. In the meantime, interest in qualitative models of space has multiplied as reflected by several recent workshops and conference sections on the subject (QUARDET 93, COSIT 93, IJCAI 93, KI-93).[1] Taking all of these new developments fully into account is impossible within the time allotted for revision, and without a complete rewrite of several parts of the text. However, I have tried to compile the references to newer published work in the bibliography, and to point to them in the appropriate sections.

I am indebted to Wilfried Brauer and Christopher Habel for their guidance

---

[1]Proceedings edited by Piera Carreté and Singh (1993), Frank and Campari (1993), Anger, Guesgen, and van Benthem (1993), and Hernández (1993c) respectively.

and support. Christian Freksa's insights inspired much of this work. The former and present members of the AI/Cognition Group provided the academic and computing environment that made this book possible. My colleagues Martin Eldracher, Margit Kinder, and Gerhard Weiß took much of the administrative burden from my shoulders during the final phase of this work. I had many fruitful discussions with Stephan Högg, Irmgard Schwarzer, and Kai Zimmermann. Daniel Kobler contributed through his own research work, through detailed proofreading of draft versions, and with clever TEX-macros to draw the icons of many tables. Viola Krings made an excellent job drawing some of the figures and tables. I want to thank Christopher Habel's group at the Universität Hamburg, particularly Simone Pribbenow, Ralf Röhrig, Rolf Sander, and Christoph Schlieder, and Josef Schneeberger (TH Darmstadt) for their critical interest in my work. Finally, I would like to express my appreciation for the hints provided by the anonymous reviewers, the help and patience of the editorial staff at Springer, and the many other people who contributed in various forms to the completion of this book.

February 1994   Daniel Hernández

Contents

Chapter 1

Introduction

*In science, one can learn the most by studying what
seems the least.*

M. Minsky, *Society of Mind, 1986.*

1.1   The relevance of spatial knowledge

The ability to cope with spatial environments is one of the fundamental survival
skills of animals all the way down to "simple" insects. As such, it has been a
central subject of study in cognitive science and artificial intelligence.

Space is a very general concept used in many different contexts to denote
different things. Freksa and Habel (1990b), for example, mention the following
four kinds of "spaces" as particularly relevant from the cognitive science point
of view: Abstract *mathematical spaces* are structures made up of arbitrary elements according to a set of axioms (Euclidean geometry is one such abstract
mathematical model, which also happens to be an axiomatization of some aspects of physical space). Concrete *physical space* is the "real" space in which
human activity evolves. It is 3-dimensional, allows only positive extension and is
accessible to objective measurement. *Psychological space* is the model of physical
space that results from perception. It is determined both by the characteristics
of physical space and the peculiarities of perception. The distinguished role of
space in human cognition is the result of the fact that space is perceived through
multiple complementary channels (visual, tactile, acoustic, etc.). *Metaphorical
space* arises from the transfer of spatial concepts to non-spatial domains such
as emotion, age, success, etc. Given the immediacy of spatial experience, this
transfer allows us to understand complex non-spatial concepts. A cognitive representation of spatial knowledge must be concerned with all of these concepts
of space. It should provide a model of physical space that is close enough to

the way it is perceived and described by humans, but at the same time can be
formalized using abstract mathematical structures. The resulting representation
should ideally also be applicable to non-spatial domains in a metaphorical sense.
We shall call such a formalizable subset of psychological space *cognitive space*.

From the artificial intelligence point of view, the representation of spatial
knowledge, particularly commonsense knowledge about space, is required for
many typical AI tasks, such as high-level vision, route planning, and physical and engineering applications. Furthermore, given the importance of spatial
knowledge in human cognition, spatial concepts play an important role in many
problem solving methods (not only those commonly considered to be graphical
or visual), and in the design of user interfaces (not only in "Graphical User Interfaces", consider common phrases such as "deeper in the system", "top level
loop" and so on).

1.2   The problem and why it should be solved

Most existing approaches to the representation of space are based on one or both
of the following implicit assumptions:

- There is an infinitesimally exact world "out there", so the more detail a
  representation contains the better.

- Computers are essentially number processors, so a numerical coordinate
  representation is the most appropriate.

Take for example the following quantitative description:

> *Rectangular box 50 units wide, 40 units high contains the following*
> *objects (positions given relative to box with origin at lower left cor-*
> *ner): object D 15 units wide, 10 units high at (35, 30); object C 6*
> *units wide, 5 units high at (39, 20); object S 17 units wide, 4 units*
> *high at (8, 36); object T 6 units wide, 16 units high at (0, 8).*

Can you figure out what it represents? It is difficult to visualize the kind of
spatial configuration described by these sentences, even if paper and pencil are
at hand. The description happens to be a straightforward verbalization of the
LaTeX-code required to draw the left floor plan in Figure 6.10 on page 113. Now
consider the following qualitative description (given to you, for example, over
the phone to tell you where to find a book):

> *When you enter the office, there is a desk in the back at the window.*
> *To its right, there is a computer table, to the right of which another*
> *desk, and an adjacent bookcase can be found. There, in the second*
> *compartment on the left, it is the third book...*

What kind of information does it contain? A context is established first by
mentioning the type of room it is supposed to be (an office), thus priming our
expectations about the objects typically found in offices. Next, the positions of

the objects in the room are described using qualitative relations like "adjacent", "to the right" and "in the back". Note that no exact statements are made, no distances or sizes mentioned. Furthermore, various levels of granularity are used. The position of the "functional group" consisting of desk, computer table, etc. in the room is described coarsely, whereas the position of the book on the shelf involves finer grained relations.

Thus, while quantitative approaches are useful in domains in which exact data is available (e.g., CAD of technical equipment) they have several drawbacks:

- Complexity: The number of values that a descriptional variable may take, affects indirectly the complexity of algorithms operating on them. Not only are more involved computations required, but the "granularity" of the representation, as determined by the fixed scale chosen, may make more distinctions than necessary for a given task. For example, the exact positions of all objects in the room are not necessary, if all we need to know is what objects are at the wall to be painted.

- Falsifying effects: Quantitative models might actually falsify the representation by forcing discrete decisions. In the quantitative description given above, we are forced to approximate the size and positions of objects to a previously fixed unit. If, for example, we were using a scale model to plan where to place furniture in a room, we might discover later, that what seemed to fit in the model does not fit in reality.[1]

- Partial and uncertain information: Due to the very nature of quantitative approaches, if a value is not known "exactly" (as determined by the a priori fixed scale), it has to be either ignored or assigned a range of possible values. Obviously this second alternative leads to increased computation. If we only know, for example, that "the bookcase is *adjacent* to the table", then various ranges of positional coordinates corresponding to the areas to each side of the table are the only way to represent that fact in a quantitative approach.

- Missing adequacy: Humans are not very good at determining exact lengths, volumes, etc., whereas they can easily perform context-dependent comparisons. Quantitative approaches, however, force us to use quantities to express even qualitative facts. Suppose, for example, we were planning where to put a desk such that "it is *close* to the window". Having to decide on an exact position for the desk, will probably make it more difficult to consider further design constraints.

- Transformational "impedance": As a consequence of the missing adequacy just mentioned, spatial reasoning systems based on numerical approaches

---

[1]This, of course, can also happen in the qualitative approach described below. The point is that a qualitative relation may be all we need to arrive at the same conclusion ("the new bookcase is *larger* than the old one"), or be even more "precise" ("the two parts have *equal* length").

might have to transform back and forth between their internal representation and one accessible to human users in order to interact with them. During these transformations, information might get lost. Continuing the previous furniture placement example, deciding on exact positions only after all design constraints have been dealt with in a qualitative way, is likely to yield better results, faster.

The representation developed in this book is intended to overcome these drawbacks, thus making systems involving spatial reasoning less complex, and more accessible to human users.

1.3   The kind of solution sought

We want to provide a model of representation that takes into account basic principles of a cognitively adequate representation of our knowledge about space, among others, the qualitativeness of spatial knowledge and its hierarchical organization, and the ability to reason at different granularity levels (fine vs. coarse reasoning). We claim that directly modeling this qualitativeness can lead to more intuitive user interfaces for applications as diverse as architectural design and Geographic Information Systems (GIS), as well as to more efficient ways to handle partial and uncertain spatial information.

We model the qualitativeness of "cognitive space" by using a *relative* representation of spatial knowledge based on *locative relations* (over selected spatial dimensions) among objects, and between objects and distinguished reference structures (e.g., landmarks and boundaries). Such a qualitative representation avoids the falsifying effects of exact geometric representations by "not committing" to all aspects of the situation being represented. By doing so, however, it is also "under-determined" in the sense that it might correspond to many "real" situations. The reason it still can be effectively used to solve spatial problems is that those problems are always embedded in a particular context, which provides the information required to complement the representation. On the other hand, if only coarse information is available, the reasoning process is less involved than if more details are known. This is a very attractive aspect of our approach, which distinguishes it from other frameworks (e.g., using value ranges or confidence intervals) in which less information means more computation.

A scene is represented as a net of mutually constraining locative relations. Inference is done by general constraint satisfaction mechanisms extended to take the structure of the domain into account, and by specialized procedures, which operate on data structures that analogically reflect the structure of the relational domain on a higher level of abstraction. The advantage of these data structures is that, since they have the same structure as the relational domain they represent, operations such as a change in point of view or the composition of relations can be performed efficiently.

Note that, because cognitive space is not directly "accessible" for study, natural language descriptions are often used as access media. It is only the relational content of those descriptions that we are interested in. That is, the linguistic issues involved in extracting spatial relations out of natural language statements are not the subject of our study. The assumption we make is that the relational content reflects the basic conceptual structures of cognitive space.

1.4 Overview

Chapter 2 introduces the concept of qualitative knowledge, contrasts it with quantitative knowledge, and discusses in general terms the properties of qualitative representations including the central role of context and structural relational domains.

Chapter 3 embeds qualitative representations in the context of an extended "Knowledge Representation Model", characterized by making the role of the observer explicit. This model gives a natural explanation for the properties of qualitative representations, and provides a uniform framework for comparison with other representation paradigms.

Chapter 4 describes a model for the qualitative representation of 2-dimensional, positional information based on topological and orientation relations. The relations are derived systematically from basic observations and their structure is made explicit. The role of reference frames if also discussed.

Chapter 5 explores the reasoning mechanisms that operate on qualitative representations to make useful inferences. Some of these are specific to the representation of spatial knowledge (e.g., transforming between reference frames, composing spatial relations, or reasoning with "abstract maps"), whereas others such as the constraint reasoning mechanisms apply to general relational representations.

Chapter 6 exemplifies the knowledge assimilation and knowledge use aspects of qualitative representations in the realm of visual processing by discussing a system to build cognitive maps from partial views and a system to visualize qualitative descriptions on screen.

Chapter 7 sketches extensions of the qualitative approach to represent positional information in 3-D scenes, as well as other spatial concepts such as size, shape, and distance.

Chapter 8 gives an overview of relevant related work in the areas of qualitative representations and the representation of spatial knowledge. It discusses the seminal papers motivating our work as well as the most important alternate approaches.

Chapter 9 contains a summary of the book, its major contributions and limitations, and an outlook on future work.

Chapter 2

Qualitativeness

> *We turn to using quantities when we can't compare the qualities of things.*
>
> M. Minsky, *Society of Mind, 1986.*

Early artificial intelligence research concentrated on modeling tasks such as chess playing or theorem proving, that demand "intelligent" conscious effort when performed by human beings. Everyday knowledge about the world, about time, space, causality, etc. seemed too obvious to deserve attention. While chess playing and theorem proving programs rapidly achieved impressive results, the kind of commonsense knowledge about the world that every child has, turned out to be stubbornly elusive to formalization.

One common trait of commonsense knowledge is its qualitative nature. When interacting with the world, people are able to handle complex situations and predict the behavior of physical objects without having to solve the kind of differential equations that a physicist would use to formally describe a physical situation. This insight has led to the very active research field of "qualitative physics". Qualitative physics models physical processes using qualitative equivalents of the tools used in traditional physical models such as variables (with small sets of possible values called "quantity spaces"), equations (relating variables, rates of change, etc.) and states (representing snapshots of the processes in time). Reasoning is done through "qualitative simulation", a process in which a sequence of discrete "episodes" is constructed, corresponding to changes in the values of qualitative variables. Curiously, most of the qualitative physics research has been concerned with modeling *processes* (Bobrow 1984; Weld and de Kleer 1990b), and not, or only indirectly, with the more basic aspects such as time and space.

In this chapter we examine the properties of qualitative representations in the context of describing spatial situations. These properties generalize to other qualitative domains as well. We start by defining the concept of qualitative knowledge, and contrasting it with quantitative knowledge. We then examine the central role of context and structural relational domains.

2.1   Qualitative vs. quantitative knowledge

The concept of "qualitativeness" is best understood in contrast with its counterpart, the concept of "quantitativeness". Typical dictionary definitions are an interesting starting point:

> **quality**: *1. The essential character of something; nature. 2. An inherent or distinguishing characteristic; property.*

> **quantity**: *1. A specified or indefinite number or amount. An exact amount or number. 2. The measurable, countable, or comparable property or aspect of a thing.*

> (excerpts from American Heritage Dictionary)

Applied to "knowledge", i.e., to what an agent knows *about* the world, this distinction means, that a qualitative representation provides mechanisms for representing only those features that are unique or essential, whereas a quantitative representation allows to represent all those values that can be expressed with respect to a predefined unit.

The measurement of a quantity implies that a number is assigned to represent its magnitude. Usually the assignment can be made by a simple comparison. The magnitude of the quantity is compared to a standard quantity, the magnitude of which is arbitrarily chosen to have the measure 1. This suggests the relationship between qualitative and quantitative knowledge: a qualitative representation can be "anchored" by establishing a correspondence between the abstract entities in the representation and the actual magnitudes. Quantitative knowledge is obtained whenever a standardized scale is used for anchoring the represented magnitudes. The use of a scale is also the context in which issues of granularity and resolution are meaningful, since a scale defines a smallest unit of possible distinction below which we are not able to say anything about a quantity.

2.2   Properties of qualitative representations

Qualitative representations have many useful properties, many of which were first pointed out by Freksa (1991). We discuss them here in general terms, while chapter 3 provides a representational framework, which explains them.

- Qualitative representations make only as many distinctions as necessary to identify objects, events, situations, etc. in a given context (recognition task)[1] as opposed to those needed to fully reconstruct a situation (reconstruction task).

- All knowledge about the physical world in general, and space in particular, is based on comparisons between magnitudes. As representations that capture such comparisons, qualitative representations reflect the relative arrangement of magnitudes, but not absolute information about magnitudes.

- Qualitative representations are "under-determined" in the sense that they might correspond to many "real" situations. The reason they still can be effectively used to solve spatial problems is that those problems are always embedded in a particular context. The context, which for simplicity can be taken to be a set of objects, should constrain the relative information enough to allow spatial reasoning, for example by making it possible to find a unique order along a descriptional dimension. In other words, a representation that can count on being used together with some particular context does not need to contain as much specific information itself.

- Qualitative representations handle vague knowledge by using coarse granularity levels, which avoid having to commit to specific values on a given dimension. With other words, the inherent "under-determination" of the representation absorbs the vagueness of our knowledge. For example, if it is unclear if the color of an object is *dark blue*, *dark brown* or *black*, a qualitative representation would record *dark*[2] as a distinction at the next coarser level of granularity. This coarse information (which, however, is *not* wrong) might be all we need to identify the object in a given context, for example, if there is only one dark object among light ones. In contrast, representation formalisms that rely on defaults[3] to handle vagueness have to record one of the specific values, i.e., they have to make the assumption that the specific value is the case (e.g., *black*). The fundamental difference between these approaches becomes evident as soon as further knowledge becomes available. While in the qualitative approach only a refinement of what is already known is needed (e.g., if the *dark* object turns out to be *dark blue*), in conventional default reasoning approaches the false assumption has to be retracted and a potentially costly revision has to be done to take back facts derived from it.

- In qualitative representations of space, the structural similarity between the representing and the represented world prevents us from violating constraints corresponding to basic properties of the represented world, which,

---

[1]The distinction between recognition and reconstruction has a long tradition in the vision literature. Johnson-Laird (1988), Freksa (1991), and others use the term 'identification' instead of 'recognition'.

[2]For the sake of simplicity, we treat here *dark* and *light* as "colors" and not as the degrees of brightness that they are.

[3]In the current AI literature, the term "default reasoning" is used to denote "patterns of inference that permit drawing conclusions suggested but not entailed by their premises" (Nutter 1987, p. 840).

in propositional systems based on defaults, would have to be restored through revision mechanisms at great cost.[4]

- Unlike quantitative representations, which require a scale to be fixed before measurements can take place, qualitative representations are independent of fixed granularities. The qualitative distinctions made may correspond to finer or coarser differences in the represented world, depending on the granularity of the knowledge available and the actual context.

- The informative content of qualitative relations varies. Some describe what would correspond to a large range of quantitative values of the same quality, while others may single out a unique distinctive value (cf. Simmons 1986).

- While the discriminating power of single qualitative relations is kept intentionally low, the interaction of several relations can lead to arbitrarily fine distinctions. If each relation is considered to represent a set of possible values, the intersections of those sets correspond to elements that satisfy all constraints simultaneously.

It should be noted that qualitative relations are not restricted to comparatives (e.g., smaller, wider) or their mathematical equivalents, partial or linear orders[5] (e.g., $\subseteq$ is a partial order on the power set, $\leq$ is a total order on the set of natural numbers). In particular, the qualitative representation of positions that we will introduce in chapter 4 uses locatives (e.g., left-back, right) and topological relations (e.g., contains, overlaps).

2.3 Structured relational domains

The search for distinctive features that characterizes the qualitative approach has an important side effect: It structures the domain according to the particular viewpoint used. Some of the qualitative distinctions being made are conceptually closer to each other than others. This structure is reflected in the set of relations used to represent the domain:

> Two relations are *neighbors* if the corresponding domain situations can result from each other without an intervening situation that could be described by a third relation of the same dimension.

This structuring of the domain has important consequences, as we will discuss in detail in later chapters. For example, neighboring relations behave similarly, which allows us to define hierarchically organized levels of granularity by considering groups of neighboring relations as a single "coarse" relation. The idea of

---

[4]However, defaults and their associated revision mechanisms are also needed in the qualitative approach at a "technical" level to handle the fact that such representations are inherently under-determined.

[5]A *partial order* is a reflexive, transitive and antisymmetric relation. A *total order* demands additionally that any two elements of the base set be comparable, i.e.,$\forall$ x,y $\in$ B, xRy $\lor$ yRx.

structured relational domains, which plays a central role in this book, has been approached from at least three different perspectives.

Hayes (1979), and later Forbus (1984), introduced the concept of "quantity spaces". A quantity space is a small, discrete set of values a physical variable may take and is usually totally ordered. Our structured domains are also small, discrete sets of spatial relations between two objects, but have a much richer structure than just total order.

Mavrovouniotis and Stephanopoulus (1988) introduced, in the context of relational representations, the related concept of "consecutive relations" to which compound order of magnitude relations are constrained. Nökel (1989) defined the concept of "convex relations" to be those between a pair of time intervals that can be transformed into one another by continuously deforming the intervals.

Finally, Freksa (1992a) introduced the notion of "conceptual neighborhood" of qualitative temporal relations, which has been the direct motivation of our work (see section 8.1).

In the context of the representation of spatial knowledge some of the basic structural properties of physical space, which have to be taken care of in a representation are:

- Physical space is homogeneous and continuous.

- Objects have only positive extension.

- Different objects cannot fill the same space at the same time (uniqueness of positions).

- Each object exists only once (identity).

- An object can only move to neighboring positions (adjacency).

Depending on the type of representation, these properties might be reflected by intrinsic properties of the representation itself (analogical aspects) or have to be modeled explicitly (propositional aspects). As we will show in later chapters, explicitly exploiting the domain structure leads to intuitive and computationally efficient representations.

Chapter 3

A cognitive perspective on knowledge representation

> *Representation is in the mind of the beholder.*
> Winograd & Flores, *Understanding Computers and Cognition, 1986.*

As the previous chapter indicated, the motivation for the qualitative approach is a cognitive one. Introspection and the way we verbalize spatial knowledge suggest that our consciously retrievable long term representation of spatial configurations is qualitative and vague. Without having to necessarily ascribe representational capabilities to the cognitive systems under study (Maturana and Varela 1980), computational models of those systems *do* need some form of representation of spatial knowledge.

In this chapter we lay down the theoretical framework for a qualitative representation of space that takes cognitive aspects into consideration. We begin with an informal assessment of the issues involved in developing a particular representation. It becomes clear that a coherent model of the representation process as a whole is required if we want to avoid getting lost in a multitude of technical issues. We present such a "Knowledge Representation Model", which is an extension of Palmer's influential framework (Palmer 1978), characterized by making the role of the observer explicit. The various modalities of representation such as declarative vs. procedural, propositional vs. pictorial, and in particular quantitative vs. qualitative, can be explained as different aspects of the mappings established by the representation model.

3.1   Issues in knowledge representation

The wide range of problems studied in the field of artificial intelligence have two things in common: The need for heuristic problem solving strategies (most of which involve search), and the need for the representation of vast amounts of knowledge. Problem solving strategies and knowledge representation are highly interdependent: The use of an appropriate representation for a given problem can greatly simplify the problem solving mechanisms needed. Historically, there has been a gradual shift from the early emphasis on search techniques to the recognition of the central role of knowledge for intelligent behavior. However, up until recently, the mainstream efforts concentrated on developing computation oriented representation formalisms and their implementations, paying little attention to the process of modeling itself.

Before considering a more comprehensive model of representation (see next section), let us look at the issues involved in developing a representation for a particular application. A good representation is well suited for a particular purpose, it exposes, for example, the inherent constraints of a problem, and facilitates its solution. Thus, the first step in the modeling process is determining the intended use of the representation. The second step is to choose the kind of entities to be represented, their attributes and the relations among them. This involves two issues: What part of the information available is relevant, and what granularity is appropriate for the task at hand. The first question is particularly important in the context of perceptual representations because of the huge amounts of sensory information available. The most accepted view is that only the information needed to discriminate an object, property or relation in a given context needs to be represented. This is the motivation for the qualitative approach to representation introduced in general terms in the previous chapter, and exemplified by the representation of positional information in 2-D in the next chapter. The second question is related to the first one, but deserves separate discussion. Suppose, for example, that you want to represent the shapes of simple 3-D objects. One possible representation is to associate a unique symbol with each type of object (e.g., block, pyramid, cylinder). The advantage of using these high-level primitives is the straightforward mapping. The disadvantage is that the procedures manipulating the representation have to deal with each type of object separately. Another possibility is to choose lower-level primitives such as a base cross-section (e.g., rectangular) together with a sweep function (e.g., linearly increasing) that generates the shape by moving the cross-section along a given axis (see section 7.4 for details). The advantage of using these low-level primitives is that more general mechanisms, that handle a wide range of similar shapes, can be applied. They also provide similarity criteria to determine similar shapes. The disadvantage is that, if the task demands handling the higher level objects, a cumbersome transformation from and to the actually used concepts has to be done repeatedly.

Because of these tradeoffs, it is generally desirable to maintain multiple levels of granularity, the issue being then how they relate.[1] In most cases, a hierarchical organization of granularity levels best reflects the natural structure of the represented domain. Complex domains are often structured taxonomically in classes and subclasses related by IS-A links. This organization allows the efficient storage of common attributes by associating them with higher level classes. Inheritance mechanisms provide then these attributes to subclasses. This leads to the distinction between terminological knowledge, capturing the structure of the domain, and assertional knowledge, expressing facts about the world using that terminology. Note that the issue of granularity in the sense discussed above is related but not identical to the quantitative vs. qualitative distinction introduced in the previous chapter. Levels of granularity occur in both quantitative and qualitative representations. The examples given above correspond to qualitative granularities. Quantitative granularities depend on the predefined scale chosen.

There are many other issues in the design of a representation that we mention here only in passing since they pertain to the representation formalism rather than to the modeling process:

- How can knowledge be accessed? Since storage and access are inverse operations, there is a tradeoff in the amount of effort invested for each of these tasks.

- How can new knowledge be inferred? That is, how can knowledge that is implied by what the system knows be made explicit?

- How can new knowledge be added to what is already known? The assimilation process can be very complicated if, for example, the new knowledge contradicts previous facts. Also, if some part of the world (or more precisely the knowledge about the world) changes, it is not always clear what happens to the rest of what the system knows (this is known as the frame problem).

- How can equality or similarity among knowledge structures be determined? (matching process)

- What kind of knowledge does the system have about its own structure? (meta-knowledge)

- If multiple representations are available, how does the system choose among them? How can consistency be maintained?

To summarize, we must distinguish between the things we want to represent and the formalisms with which we represent them. The important issues arise in the process of selecting and structuring what we want to represent, i.e., in the modeling process, which is independent from the particular formalisms in the

---

[1]A related issue is, if there exists a "canonical representation". This is generally a granularity level from and to which various other levels or formats can be easily transformed, or which otherwise allows a uniform manipulation of the knowledge represented.

same way that an abstract specification of a procedure is independent from its concrete realization in a programming language.

3.2   Knowledge representation model

The kind of comprehensive view of representation that takes the modeling process into consideration has been proposed by several authors. In an early and influential paper, Bobrow (1975) suggested viewing representations as the result of a selective mapping of aspects of the world. He posed the central questions about domain, range, and operational correspondance in a representation:

- What is being represented?

- How do objects and relationships in the world correspond to units and relations in the model?

- In what ways do the operations in the representation correspond to actions in the world?

That only "aspects" of the world can be represented is stressed by Marr (1982), who defines representation as a "formal system for making explicit certain entities or types of information, together with a specification of how the system does this." Making certain aspects explicit facilitates the solution of particular tasks, while making it harder to recover other types of information that are pushed in the background. Marr's now classical example is the representation of numbers using the Arabic, Roman, and binary numeral systems: The representation consists of strings of symbols from the corresponding alphabets and rules for constructing the descriptions of particular integers. In the Arabic system the integer is decomposed into a sum of multiples of powers of 10. Thus, while the Arabic representation makes the decomposition into powers of 10 explicit (making it easy to determine if a number, say 10000, is a power of 10), the binary representation makes the factorization into powers of two obvious. Furthermore, operations such as addition and multiplication are easy if the Arabic or binary representations are used, but not that easy with Roman numerals (particularly multiplication).

As a first step toward the goal of understanding the nature of cognitive—particularly, perceptual—representations, Palmer (1978) developed a general representation theory, based on the distinction between "two related but functionally separate worlds: the *represented* world and the *representing* world." He enumerates the elements required to fully specify a representational system as follows:

1. What is the represented world?

2. What is the representing world?

3. What aspects of the represented world are being modeled?

4. What aspects of the representing world are doing the modeling?

5. What are the correspondences between the two worlds?

The task for which the representation is intended must have been specified previously in order to answer these questions.

Furbach et al. (1985) made the object and structure defining relations implicit in 3. and 4. above more explicit by introducing the concept of a *reference world* associated with each of the represented and the representing worlds. Which aspect of the represented world is being modeled by which aspect of the representing world is determined by the object and structure defining relations comprised in their corresponding reference worlds and by the *mapping* between them. Rehkämper (1991) further analyses these "reference worlds", distinguishing four corresponding components in each of them: Objects, properties of objects, relations among objects, and operations on objects.

As Palmer points out, the concept of representation is not a static one. There is a deep dependence of representation on processing, because processing operations functionally determine the relations that hold among the object elements. For example, in a scheme in which the relative sizes of objects are represented by explicit pointers between nodes corresponding to those objects, the operations to find nodes and to follow arcs are an integral part of the representation. For, regardless of what other information might be implicitly available in the representing structures, the only actually useful knowledge is that for which interpretative operations are defined.

We advocate an even more radical view of the semantics of representations based on the observation made by Winograd and Flores (1986), that the essence of computation is the correspondence between manipulation of formal tokens and the attribution of meaning to those tokens by somebody. That is, not the internal interpretative operations nor the mapping between the represented and representing worlds alone, establish the meaning of a representation, but the association of these elements established in the mind of the entity that designs or uses the representation. We call this entity the *observer*, and make it an explicit part of the representation model. The observer also plays a central role in the modeling process leading to the design of a representation, for objects, properties, relations, and operations are *relevant* only in the domain of distinctions made by an observer.

We already mentioned a further sense in which operations play an important role. One of the reasons for constructing representations is being able to foresee the consequences of actions in the represented world by manipulating the representing world. Therefore, we have to establish a correspondance between actions on objects of the world being represented and operations on the representing elements. This assumes a view of the world in which actions produce changes that lead from one state of the world to another. An alternate view, which might be more appropriate in some applications such as the qualitative simulation of physical phenomena (Forbus 1984), is to concentrate on processes in the world and their direct models in the representation.

Palmer distinguishes *non-equivalent* representations, in which different relations are modeled, from *informationally equivalent* representations, in which the same relations are modeled, but in different ways. In analyzing the ways in which representations of the world can differ, he defines some of the basic concepts in representation theory: A *dimension* is a set of mutually exclusive relations. Examples of dimensions are height and distance. The *resolution* is the number of relations in a dimension. This is related to the granularity issue discussed in the previous section in its most basic form, i.e., without taking structure into consideration. He distinguishes *nominal* dimensions, where it can only be determined if two relations are different or the same (i.e., only identity is preserved), from *ordinal* dimensions where the ordering of relations is preserved. Palmer makes this distinction with very simple dimensions in mind.[2] As we showed in section 2.3, relational domains can have a more involved structure than just a linear order. Finding out that structure, and constructing representations that preserve it, is an essential aspect of the approach advocated in this book.

As an aside, it is interesting to look at the sign concept in linguistics as originally introduced by de Saussure (1916), and its relation to the knowledge representation model presented here. For de Saussure, a sign is a bilateral entity consisting of an external form (the signifier or *signifiant*) and an associated concept (the signified or *signifié*). This association is fixed but arbitrary in any given language. The signified is a conceptual entity and should not be confused with the external object (*chose réelle*) to which it refers. In particular, the actual meaning of a concept depends not only on its relation to an external entity, but also on what other concepts exist. That is, the structure of language as an immanent system of mutually dependent relations is the central concept. The parallel to the knowledge representation model is evident. If we identify the represented world with the external objects and the representing world with the external form, the association between these two as provided by the system of concepts corresponds to the attribution of meaning in the mind of the observer. The idea that concepts are not isolated entities but are part of a structured complex of mutually dependent relations is a very powerful one. As we will see in section 4.4, for example, the relation `back` describes different relative orientations depending on the other distinctions made at its level of granularity (e.g., {`back`, `front`} vs. {`back`, `front`, `left`, `right`}). Furthermore, in the area of knowledge acquisition (and its "extrapolation", machine learning) this leads to the insight, that we (or for that matter, the artificial systems we design) can "know" only relative to what we already know. This does not, however, imply innate knowledge, but a gradual concept formation strategy, where new concepts are first expressed in terms of old ones until they become part of the network of mutual dependencies (and thus, by the way, change the value of other previously known concepts).

Figure 3.1 summarizes the knowledge representation model introduced in this section. The object- and structure defining relations of the reference world

---

[2]He does, however, also consider inter-dimensional structure and argues that it, too, has to be preserved in the representing world for the modeled dimensions.

R_1 impose a particular view on the represented world W_1 (i.e., a set of objects, properties of objects, relations among objects, and operations on objects). The step from W_1 to R_1 involves the modeling process discussed in section 3.1. The representing world W_2 is the medium of representation, of which only the aspects defined by the reference world R_2 are actually used. The step from R_2 to W_2 involves the implementation of a representation. The mapping C between R_1 and R_2 leads to the attribution of meaning to the representation by an observer. Note that this mapping is not necessarily (not even usually) type preserving, that is, objects of R_1 may be mapped to relations in R_2, and vice versa.

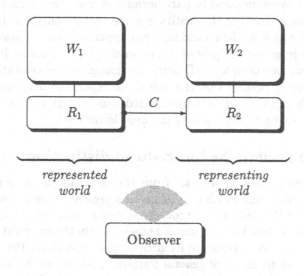

Figure 3.1: Knowledge Representation Model

W_1 represented *world*; W_2 representing *world*; R_1, R_2 corresponding *reference worlds*; C mapping between selected aspects of the represented and the representing worlds as defined by R_1 and R_2

3.3 Modalities of representation

Many heated debates in the field have centered on modalities of representation characterized by concept pairs such as declarative vs. procedural, propositional vs. analogical,[3] and so on. In the context of the knowledge representation model introduced in the previous section, these modalities correspond to different kinds of possible mappings between the represented and the representing worlds.

---

[3]The famous imagery debate about the nature of perceptual representations contrasts 'propositional' with 'pictorial'. Pictorial representations are a special case of analogical representations, which are the more appropriate counterparts of propositional representations.

The goal of this section is to embed qualitative representations in the context of other modalities of representation, and thus answer questions such as: In what sense are qualitative representations different from general relational representations? Can qualitative representations be analogical at the same time? Are relative representations the same as qualitative ones?

It must be said from the outset, that the modalities of representation discussed in this section do not exist in "pure" form. A typical representation rather has some propositional aspects, some analogical aspects, and so on, whereby one of these aspects might be the distinctive one. For example, many notations such as musical scores and chemical formulae convey information in part because of propositional conventions, and in part because of analogical structure. Furthermore, in the many levels that constitute a representation, from its conceptual structure to its actual implementation, what appears to be declarative or propositional at one level, can be procedural or analogical at another level, and vice versa. Consider, for example, a 2D-array as analogical representation of a chess board (neighboring positions on the board correspond to neighboring array elements), which in turn can be implemented non-analogically as a vector with appropriate indexing mechanisms at the next lower level.

3.3.1   The declarative/procedural distinction

Declarative representations abstract from the structure of the represented domain, an aspect they share with propositional representations as discussed below. In the context of the declarative/procedural distinction, declarative representations stress the explicit formulation of knowledge about the world (i.e., objects, properties and relations) separate from the knowledge about the operations on them, allowing us to use very general-purpose mechanisms for inference. Procedural representations, in contrast, represent by virtue of the operations performed by a program. There is an inherent duality in this distinction, similar to the program/data duality. Programs operate on data, but are themselves data to the compilers or interpreters that perform them. In the same sense a declarative representation on one level can be seen as a procedural specification on another (pure PROLOG being the typical example for this view).

3.3.2   The propositional/analogical distinction

The propositional/analogical[4] distinction is a more fundamental one. In an earlier paper, Palmer (1975b) proposed analyzing first a series of more basic concept pairs to get at the core of that distinction. Among the concept pairs he analyzes in the context of perceptual representations, the most relevant to our discussion are the following:

Implicit vs. explicit: The question of whether information is represented implicitly or explicitly often leads to confusions, because the concepts are used

---

[4]Following Sloman (1971), we prefer the term "analogical" instead of "analog" (which is often used synonymously in the literature) to avoid confusion with analog as "continuous" (e.g., analog data, analog device).

in different, and even opposite senses in the literature. In one sense, pictorial representations are considered "to allow the explicit representation and direct retrieval of information that can be represented only implicitly in other types of representations and then has to be computed, sometimes at great cost, to make it explicit for use" (Chandrasekaran and Narayanan 1992). In the sense preferred in this book, the implicit/explicit distinction refers to the way the represented and the representing domains are related. Thus, while the structural properties of space must be explicitly stated in a propositional representation, they are implicitly given by the corresponding properties of the pictorial representations. We use the terms implicit and intrinsic, on the one hand, and explicit and extrinsic, on the other, synonymously, except in section 4.4, where, for technical reasons, we distinguish between explicit, implicit, intrinsic, and extrinsic frames of reference.

Holistic vs. compositional: A holistic representation can only be understood as a whole and not as sum of its parts, whereas what a compositional representation represents can be derived systematically from what its parts represent. To equate holistic with analogical and compositional with propositional, as is often done in the literature, is wrong. They both can be used to represent things that have parts and relations between parts. It depends on the level at which the properties are seen, and how the mapping between represented and representing reference worlds is done. Thus, while there are emergent properties such as closedness, area, and symmetry that arise when three lines are arranged to form a triangle (and that are not properties of the component lines), analogical representations can have meaningful parts. Similarly, even though information is typically encoded componentially in a propositional representation, some properties such as the global inconsistency of pairwise consistent constraints arise only when the set of constraints is seen as a whole.

Absolute vs. relative: In the strict physical sense, an absolute representation uses a reference framework derived from fundamental relationships of space, mass, and time. In an informal sense, every representation that uses a predefined, fixed reference framework is called absolute. A relative representation establishes context dependent relations among represented entities based on comparisons between magnitudes. Again, the propositional/analogical issue is independent from this distinction.

In the sense sketched in this section, propositional representations are not restricted to propositional logic. Palmer defines propositional representations as "those in which there exist relational elements that model relations by virtue of themselves being related to object elements" (Palmer 1978, p.294). Thus, all forms of sentential[5] representations including natural language and programming languages are covered by this characterization. Similarly, analogical representations are not limited to pictorial or diagrammatical ones. They include all those,

---

[5]Sloman (1971, 1975) calls them "Fregean".

where basic properties of the represented domain result from inherent properties of the representing domain (Sloman 1975). This requires viewing the data
structures and the mechanisms operating on them as a whole when establishing
the similarity preserving mapping.

The key to understanding the propositional/analogical distinction lies in the
way the structure of the represented world is mapped to the representing world.
In the case of propositional representations, the representing world has no structure of its own (at least none of relevance to the representation). Thus, whatever
structural properties it represents have to be explicitly expressed, and result from
the mapping from the represented world. In the case of analogical representations, the structural aspects of the represented world are represented by the
inherent structure of the representing world.

3.3.3   The qualitative/quantitative distinction

This distinction was introduced at length in the previous chapter. In this section,
we want to discuss it in the context of the knowledge representation model, and
contrast it with the other modalities described above.

While an explanation of the other modalities in the context of the representation model was given without explicit reference to the observer, he plays a key
role in the characterization of the qualitative/quantitative distinction. It is only
when we make the observer explicit that the definitions of qualitative representations given make sense: "making only relevant distinctions" requires someone
or something (task) to determine what is relevant or not in a given context.
Once the central role of the observer has been recognized, however, it is only
natural to try to find a representational mapping that is as close as possible to
his conceptual categories (as opposed to one that is as close as possible to the
representational structures that happen to be available).

As far as its relation to the propositional/analogical distinction is concerned,
the typical claims are that analogical representations are quantitative while
propositional representations are qualitative. This, however, is not the case.
The concept pairs are "orthogonal" in the sense that they pertain to different
aspects of the representation process. The quantitative/qualitative distinction
has to do with the way the represented world is modeled, i.e., the contents of
the reference world. The propositional/analogical distinction has to do with
the way the represented reference world is mapped to the representing reference
world. Thus, propositional representations can be quantitative and analogical
representations can be qualitative, and vice versa (see also section 5.4.2). In
the 2-D representation introduced in the next chapter, for example, while the
distinctions made are expressed propositionally in form of relations, the data
structures used analogically reflect the structure of the relational domain on a
higher level of abstraction.

There is another sense in which qualitative representations relate to a special
form of analogical representations: diagrammatical ones. In the narrow sense
of sketch-like representations (Funt 1980; Mackworth 1977b) diagrammatical
representations are qualitative in nature, since they:

- make only as many commitments as necessary;

- are under-determined (i.e., stand for a whole class of possible instances) and context dependent;

- allow reasoning at various levels of granularity (coarse vs. fine reasoning).

3.4 Summary

Knowledge representation is at the heart of every artificial intelligence system. The issues involved in the design of a representation are not restricted to technical aspects of the representation formalism (storage, access, inference, assimilation, consistency), but include the modeling process (relevance and granularity issues). We propose an extended knowledge representation model, characterized by making the role of the observer explicit. We use the representation model to look into the various modalities of representation such as declarative, procedural, propositional, analogical, etc. Generally only some aspects of a representation correspond to a particular modality, depending on the level of abstraction considered. The concept of qualitativeness is found to be orthogonal to the modalities discussed.

Chapter 4

Qualitative representation of positions in 2-D

> *If you were walking through a circular tube, you could scarcely keep from thinking in terms of bottom and top and sides—however vaguely their boundaries are defined. Without a way to represent the scene in terms of familiar parts, you'd have no well-established thinking skills to apply to it.*
>
> M. Minsky, *Society of Mind, 1986.*

Allen (1983) introduced an interval-based temporal logic, in which knowledge about time is maintained qualitatively by storing comparative relations between intervals. The elegance and simplicity of that approach has inspired several efforts to extend it to spatial dimensions. One way this has been done is by making the same kind of distinctions as in the temporal case for the two axes of a Cartesian coordinate system. Guesgen (1989), for example, uses the relations *left of*, *attached to*, *overlapping*, *inside*, and their converses. Object boundaries are then projected onto the two axes and a pair [x-Relation, y-Relation] is used to give the relative position of objects. This approach, however, lacks "cognitive plausibility": People don't walk around decomposing the world into axes and then determining a qualitative relation for an interval on each of them![1]

This chapter introduces a model for the qualitative representation of positions in 2-D space. We chose 2-D positional information to demonstrate the usefulness of the qualitative approach in the spatial domain, because it is the simplest kind of spatial information of practical relevance. The 1-D spatial case, while different from the temporal case, can be handled by a simple modification of Allen's calculus. Extensions to other spatial aspects will be presented

---

[1]The drawbacks of this kind of extensions are discussed in further detail in section 8.1.

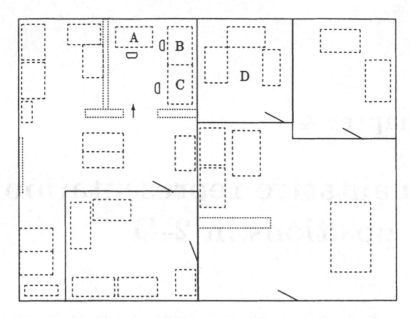

Figure 4.1: Example of 2-D projection of a scene: Layout plan
of an office

in chapter 7. We begin by characterizing the type of scenes we will be considering (2-D projections of 3-D scenes). We then explore the type of dimensions
required for their representation and present a declarative syntax for expressing
those dimensions.

4.1   2-D scenes

The type of spatial scenes that we want to model are *2-D projections* of 3-D
scenes. 2-D projections are the simplest way to depict spatial configurations,
yet they are powerful enough to allow for interesting applications such as architectural design (see for example the layout plan of an office in Figure 4.1). There
are several issues that need to be discussed when focusing on projections.

4.1.1   Characteristics

Collapsing three dimensions into two implies the possibility of objects "overlapping" each other or even being "equal" (i.e., having the same projection). Thus
projections do not have the "uniqueness of positions" property of space introduced in section 2.3. Two projections can partially or totally occupy the same
space at the same time, and it is not possible to distinguish which object is above
the other one. As a means of representing 2-D *positions*, however, this fact has
the useful side effect of providing a hierarchical organization of space. All objects

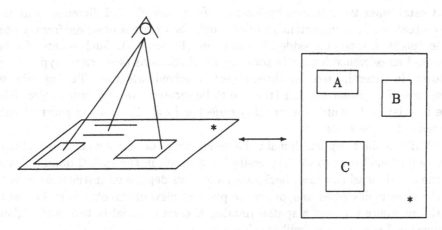

Figure 4.2: External observer vs. embedded point of view

included in another one can be manipulated at a higher level by manipulating the parent object. For example, all the objects in a room inherit the orientation of that room with respect to others in the building. Furthermore, point set topology provides the appropriate reasoning tools to handle overlapping areas. Yet, in some applications in which only solid objects occur (such as blocks in document layouts—see section 8.1), ruling out overlap and containment greatly simplifies the reasoning process.

The shape of objects and consequently of their projections also affects the complexity of the reasoning process. Some initial thoughts on the qualitative description of shapes will be presented in section 7.4. Here, however, we ignore shape by considering only either "convex" objects or canonically oriented "delineative rectangles" enclosing the objects to be represented. Also, objects are assumed to have no holes and be embedded in a plane (i.e., a space such as the closed surface of a sphere is not allowed).

These restrictions are made "for technical reasons", in order to be able to apply point set topology methods for formal reasoning. Their impact on the practical applications of the approach is not as large as might be thought. For example, lightly concave objects (in the sense that no other object in the scene can be considered to be in the cavity of another object), and objects with small holes can be tolerated, because they do not affect the position of the object as a whole. The size proportions of objects influence the type of positional relations that can hold between them. For example, large elongated objects and small compact objects tend to be in one of the coarse back/front or left/right relations to each other. For some applications it is useful to restrict size proportions to a small factor. This will be explored in further detail in section 4.4.

Determining relations in an n-dimensional space requires an "external" observer in the $(n+1)$st-dimension. This is also true for $n = 3$, since the passage of time, which constitutes the fourth dimension, is necessary to establish relations in 3-D space. For the 2-D case at hand, we assume an external observer that establishes the relations by looking "from above" (3rd dimension) at the 2-D situation. It is important to distinguish the external observer from a possible "point of view" embedded in the scene (Figure 4.2). Such points of view are used as reference frames in some types of relations (see deictic type of use below). In general, there are three objects involved, when establishing relative positions: The *primary object* (the one to be located), the *reference object* (the one in relation to which the primary object is located), and the *point of view* embedded in the scene.

While in the temporal domain the "beginning" of an interval comes always before its "end" (due to the irreversibility of time), in the spatial domain (even in the one-dimensional case) beginnings and ends depend on intrinsic properties of the objects described and/or on the point of view of the observer. The issue of the reference frame of a spatial relation becomes crucial in two- and higher-dimensional spaces, as we will see shortly.

4.1.2   Relevant dimensions

In chapter 2 we defined qualitative representations as those based on direct comparisons between objects and making only as many distinctions as necessary. The question here is what distinctions are necessary to qualitatively describe *positions* in 2-D.

As it turns out, two factors determine the relative position of objects in 2-D space: the relative orientation of objects to each other and the extension of the involved objects. Considering these factors independently from each other results in two classes of spatial relations:

- topological relations (ignore orientation)

- orientation relations (ignore extension, i.e., objects = points)

Our goal is to combine these two classes of relations to provide a model of orientation that accounts for extended objects. For this purpose we define a small set of spatial relations from the two relevant dimensions topology and orientation. Topological[2] relations describe how the boundaries of the two objects relate. Orientation relations describe where the objects are placed relative to one another. The relative position is given by a topological/orientation relation pair.

Although one of the motivations for the qualitative approach is to capture the kind of spatial information contained in natural language descriptions, the relations should not be equated with the spatial prepositions they are related to. Whenever possible, the names of the relations were chosen purposely not to coincide with spatial prepositions, because there is more to the use of prepositions in natural language than what a straightforward mapping would suggest.

---

[2]Previous work used the term "projection" instead of topological. This has been changed because of the possible confusion with the use of the word in "2-D projections of 3-D scenes" and in "projective spatial prepositions" (which are related to what we call orientation relations).

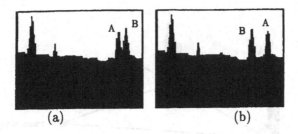

Figure 4.3: Two views of a city skyline (Handout, Schlieder 1991)

To make our point, let us take a brief look at the use of spatial prepositions in natural language, in particular prepositions used to describe the location of one object in relation to another, before studying the abstract relations in the following sections. They can be classified in topological prepositions (e.g., *near*, *at*, *in*, *on*, *between*) and projective prepositions, which in turn form a "primary deixis system" (*here*, *there*) and a "secondary deixis system" (*in front of*, *behind*, *left of*, *right of*, *beside*). Furthermore, there are prepositions describing the vertical axis (*over*, *under*, *above*, *below*, *on top of*, *on bottom of*, *underneath*, *higher*, *lower*), prepositions establishing relations among several objects (e.g., *between*, a relation among three objects), and even prepositions describing non-adjacent positions, such as *besides* (leaves open, which of two opposite positions is the case). This list omits dynamic uses of prepositions and verbs conveying spatial information.

In the next section, we first discuss arrangement as the simplest form of positional information. We do so, even though our model does not use arrangement information directly, because it exemplifies important properties of a qualitative representation and serves as the foundation of the orientation dimension. Topological and orientation relations are described in detail in later sections.

4.2 Arrangement

The simplest form of qualitative positional information is given by what is called "arrangement" (Schlieder 1990a, 1990b). *Arrangement* is the order in which landmarks appear on a view when scanning it from, say, left to right from an embedded point of view. Figure 4.3 (taken, as all other figures in this section, from Schlieder's papers) shows two views of a city skyline from different points of view. A *view* can be more generally defined as the sequence in which a sweep line centered on the point of view touches the landmarks in a scene when rotating 180° clockwise.[3]

---

[3]Distinguishing left and right in a view assumes a maximal view angle of 180°. Furthermore, landmarks are assumed to have no extension (i.e., to be points) and be visible (no occlusion).

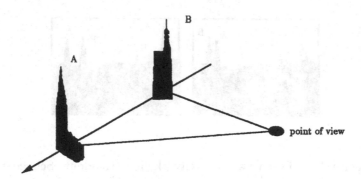

Figure 4.4: Triangle orientation (Handout, Schlieder 1991)

Arrangement provides rough information on the location of the point of view: in Figure 4.3a since A is left of B, the point of view must be on the left side of the directed straight line connecting the two landmarks A and B (see also Figure 4.4). Equivalently, the point of view is on the left side, if visiting A, B and C in sequence implies a positive (counterclockwise) rotation. Thus the "orientation"[4] of the triangles formed by landmarks and point of view describes their positions.

The question arises, how the local arrangement views relate to the corresponding global view "from above". To answer it, a relationship must be established between views and "locations", which are defined below. For every pair of landmarks there is a directed straight line connecting them. This line is not necessarily unique, for two or more landmarks might be collinear, i.e., aligned on the same line. A *location* can be characterized by stating for each of the lines if it is on their left side (+), their right side (−) or on the line itself (0). This information is summarized by a *position vector* $p(S) = (g_1(S), \ldots, g_m(S))$ where $g_i(S)$ is a function yielding +, −, 0 depending on how the point S relates to the line i. There are three types of locations (examples refer to Figure 4.5):

points,   for example $p(S_0)$ =(-0+--0+++++-0+-++)
segments,   for example $p(S_1)$ =(0+++++++++++--+---)
areas,   for example $p(S_2)$ =(-++---++-+-----++)

All points of a location have the same position vector, which is also unique for each location. However, a view does not describe a location uniquely. It is necessary to keep track additionally of which part (left of right) of the sweep line touched the landmark first. For example, the sequence BCdFEaG uniquely characterizes the corresponding location (where uppercase letters are used if the left part, lowercase letters if the right part of the sweep line touched the corresponding landmark first). Such a sequence is called a *panorama* (see Figure 4.6).

---

[4] *Orientation* is used here in the restricted technical sense of the direction of rotation needed to visit the nodes in a given order, and should not be confused with the orientation dimension introduced in section 4.4. It is negative (−) for a clockwise rotation and positive (+) for a counterclockwise rotation.

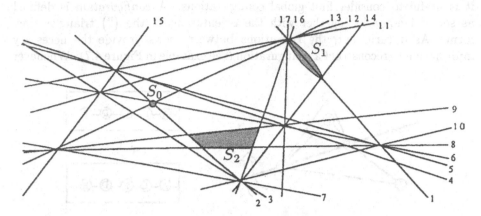

Figure 4.5: Three types of locations (Figure 5 from Schlieder 1990a, p. 135)

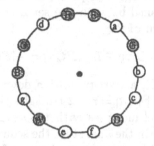

Figure 4.6: Panorama for the given example (Figure 12 from Schlieder 1990a, p. 140)

By the above construction, every location has a unique panorama. Conversely, if a panorama is given, the position vector of the corresponding location can be computed as follows:

1. Extract all views of a location from the panorama: Start with all uppercase letters as the first view, e.g., [BCFEG] and then strip successively the first one to form the next view, e.g., [CFEG], [FEG]. Whenever a lowercase letter occurs, however, use it as the last *uppercase* letter on the next view, e.g., [FEGD]. This follows from the sweep procedure used to obtain the panorama.

2. If there exist one view in which P is to the left of Q, then the location is to the left of the directed line PQ, otherwise to the right.

The above procedure provides a method for recovering the local views out of the global panorama. To reconstruct the global panorama out of local views it is useful to consider first global configurations. A *configuration* is defined as set of landmarks together with the orientations of the $\binom{n}{3}$ triangles they form. As it turns out, the transitions between views provide the necessary information to reconstruct a configuration (see example in Figure 4.7): Whenever

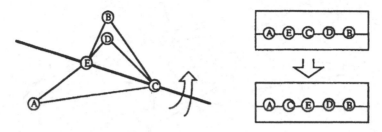

Figure 4.7: Transitions imply orientations (Figure 3 from Schlieder 1990b, p. 163)

two landmarks P and Q exchange their places in consecutive views, the following relationship can be established between the set L of landmarks left of PQ and the set R of landmarks right of PQ:

$$\forall x \in L, \forall y \in R : [PQx] = -[PQy]$$

Thus, if n points are visible, a view transition determines the relationship between the orientations of $|L| + |R| = n - 2$ triangles. In the worst case, however, the number of views needed increases with the square of the number of landmarks in the configuration. In the final step, the sequence of landmarks (or their complements) making up the panorama is obtained from the orientations of the configuration as explained in further detail in Schlieder (1990b).

Arrangement information has several of the desirable properties of qualitative representations:

- It abstracts from metric information and is thus capable of representing the kind of under-determined information typical in verbal descriptions of space. (Note, however, that it is also not "topological", since it depends heavily on straight lines, which are not preserved under arbitrary topological transformations.)

- Given an appropriate number of landmarks, it is possible to restrict the position being described to arbitrary "precision". With other words, it is independent of a given fixed granularity of representation.

Arrangement information is very common in panoramic view descriptions of the kind found in hiking books. Some of the core texts of the well known LILOG-Project are of this kind (Habel and Pribbenow 1988). Arrangement information has also been used for robot navigation by Kuipers and Levitt (1988), and Levitt and Lawton (1990) based on the same fundamental observation relating connecting lines between landmarks and the position of the point of view mentioned above (see also section 8.2.2). Navigational subgoals are given in terms of crossing lines connecting visible landmarks. This is vague enough to allow the robot to avoid local obstacles, while providing enough information to find the global goal. For this kind of applications, the simplicity and mathematical soundness of arrangement provides a useful source of information. For others, however, it is too low level, because all objects have to be related to each other on an all-to-all basis. As a consequence, the complexity of algorithms operating on configurations is exponential in the number of objects. Furthermore, it is not always feasible to construct straight lines and perform the sweep procedures required for reasoning with arrangement information alone. Particularly in small scale spaces, more informative relations such as "left-back" can be established, as we will see in the next two sections.

4.3 Topological relations

If we disregard the relative orientation of objects to each other, we can still distinguish various ways in which their projections can be related to each other. A first set of distinctions that comes to mind on how objects and consequently their projections relate is the following:

- they are *far* apart from each other,

- they are *close* together, but do not touch,

- they *touch*,

- they *overlap*,

- one of them is *included* in the other.

Although this seems to be a reasonable set of relations to start with, it turns out to have several drawbacks.[5] The first distinction between objects close together or far apart cannot be defined by topological means, i.e., by comparing just the two objects involved. It requires an external frame with respect to which the positions of the objects being compared must be established first. We shall see later that the relative size of objects provides one such external criterion allowing a coarse form of far/close distinction. For now, however, we rule out distance, since it is not preserved by topological equivalence. A further representational concern is that the set of distinctions be complete, in the sense that it covers all possible situations. Also, the corresponding relations should be mutually exclusive. Thus, each possible situation corresponds to one and only one relation. The given set of relations does not fulfill these requirements. There is, for example, no way of describing two projections being equal.

---

[5]We will discuss other approaches found in the literature (and their drawbacks) in section 8.1.

4.3.1   Systematic derivation of topological relations

What is needed is a systematic way of describing topological relations. Set theory is a good starting point: The projections are given by the sets of points they consist of, and the relations are defined in terms of set operations. Güting (1988), for example, gives the following definitions in terms of the set operations $=, \neq, \subseteq, \cap$:

$$
\begin{aligned}
x = y &:= \text{points}(x) = \text{points}(y) \\
x \neq y &:= \text{points}(x) \neq \text{points}(y) \\
x \text{ inside } y &:= \text{points}(x) \subseteq \text{points}(y) \\
x \text{ outside } y &:= \text{points}(x) \cap \text{points}(y) = \emptyset \\
x \text{ intersects } y &:= \text{points}(x) \cap \text{points}(y) \neq \emptyset
\end{aligned}
$$

As Egenhofer and Franzosa (1991) point out, this set of relations is neither complete nor mutually exclusive. As defined, *equal* and *inside* are special cases of *intersects*. There is also no way of distinguishing *touches* from *intersects* (overlaps), because that distinction is based on the difference between the boundary and the interior of point sets, which is not made here.

That observation led Egenhofer and Franzosa (1991) to the definition of topological relations based on the four intersections of the boundaries (∂) and interiors ($^\circ$) of two sets A and B.[6] The *interior* of a set A is the union of all open sets in A. The *boundary* of a set A is the intersection of the closure of A and the closure of the complement of A. The *closure* of A is the intersection of all closed sets of A. The *complement* of A w.r.t. the embedding space is the set of all points of the embedding space not contained in A. Since an intersection can be either empty ($\emptyset$) or not empty ($\neg\emptyset$), for the four intersections $\partial A \cap \partial B$, $A^\circ \cap B^\circ$, $\partial A \cap B^\circ$, and $A^\circ \cap \partial B$ we obtain $2^4 = 16$ combinatorially possible relations (Table 4.1).[7] Figure 4.8 shows geometrical interpretations of those 16 relations. Obviously, some of these relations do not have meaningful physical interpretations. For example, r_2 implies projections without boundaries, r_9 requires one of the objects to have a boundary but no interior, and so on. Imposing the restrictions of physical space, we can in fact eliminate half of them (see Table 4.2). An alternate derivation of the set of topological relations based on how regions can 'connect' to each other has been developed by Cohn, Cui, and Randell (see section 8.1.3). Selecting the subset of mutually exclusive and pairwise disjoint relations from the resulting set yields the same set of eight topological relations shown above.

4.3.2   Properties of the derived set

The remaining eight relations

---

[6]An extended '9-intersection' model is proposed in Egenhofer (1991)—see section 5.2.1. More recent developments of this framework are described in Egenhofer and Al-Taha (1992), Egenhofer and Sharma (1992, 1993).

[7]Wazinski (1993) proposes alternatively to use "graduated" topological relations based on the degree of ovelap of the surfaces involved. This, however, requires metric information about those surfaces.

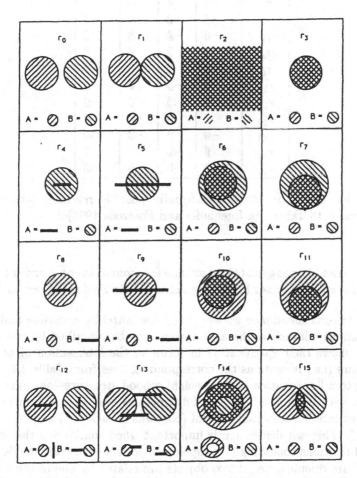

Figure 4.8: Geometrical interpretations of binary topological relations (Figure 6 from Egenhofer and Franzosa 1991)

$X\varphi Y$	$\partial \cap \partial$	$°\cap°$	$\partial\cap°$	$°\cap\partial$
r_0	$\emptyset$	$\emptyset$	$\emptyset$	$\emptyset$
r_1	$\neg\emptyset$	$\emptyset$	$\emptyset$	$\emptyset$
r_2	$\emptyset$	$\neg\emptyset$	$\emptyset$	$\emptyset$
r_3	$\neg\emptyset$	$\neg\emptyset$	$\emptyset$	$\emptyset$
r_4	$\emptyset$	$\emptyset$	$\neg\emptyset$	$\emptyset$
r_5	$\neg\emptyset$	$\emptyset$	$\neg\emptyset$	$\emptyset$
r_6	$\emptyset$	$\neg\emptyset$	$\neg\emptyset$	$\emptyset$
r_7	$\neg\emptyset$	$\neg\emptyset$	$\neg\emptyset$	$\emptyset$
r_8	$\emptyset$	$\emptyset$	$\emptyset$	$\neg\emptyset$
r_9	$\neg\emptyset$	$\emptyset$	$\emptyset$	$\neg\emptyset$
r_{10}	$\emptyset$	$\neg\emptyset$	$\emptyset$	$\neg\emptyset$
r_{11}	$\neg\emptyset$	$\neg\emptyset$	$\emptyset$	$\neg\emptyset$
r_{12}	$\emptyset$	$\emptyset$	$\neg\emptyset$	$\neg\emptyset$
r_{13}	$\neg\emptyset$	$\emptyset$	$\neg\emptyset$	$\neg\emptyset$
r_{14}	$\emptyset$	$\neg\emptyset$	$\neg\emptyset$	$\neg\emptyset$
r_{15}	$\neg\emptyset$	$\neg\emptyset$	$\neg\emptyset$	$\neg\emptyset$

Table 4.1: The 16 combinatorially possible relations (corresponds to Table 1 in Egenhofer and Franzosa 1991)

disjoint (d), tangent (t), overlaps (o), contains-at-border (c©b), included-at-border (i©b), contains (c), included (i), equal (=)

fulfill the criteria mentioned above: They are mutually exclusive and all possible situations allowed by the assumptions can be described by one of them. Table 4.2 shows their specification in terms of the intersection of boundaries and interiors (i.e., it contains the corresponding lines from Table 4.1, arranged according to a linearization of their neighborhood structure—see next section). Compared with the first ad hoc set of distinctions, we can now express equality and differentiate between containment (contains or included) and "containment at border". This last distinction is important when considering the compound positional relations consisting of a topological/orientation pair: While {d, t, o, c©b, i©b} are oriented, i.e., if two objects are related by one of these relations it is possible to establish their relative orientations as well, this is not the case for {c, i, =}, which preclude orientation. We disallow orientations for the c and i relations to avoid having to distinguish *inside* and *outside* orientations as in the case of spatial prepositions, where "front" can mean opposite directions depending on whether the primary object is located *inside* (for example in the sentence "The altar is in the front part of the church") or *outside* (for example, "The fountain is in front of the church") the reference object. We assume the "outside" meaning of orientation, and use containment as a means of hierarchical decomposition of space, that is, the "containing" object is used as parent object and not as reference object.

$X \varphi Y$		$\partial \cap \partial$	$° \cap °$	$\partial \cap °$	$° \cap \partial$
r_0	disjoint	$\emptyset$	$\emptyset$	$\emptyset$	$\emptyset$
r_1	tangent	$\neg\emptyset$	$\emptyset$	$\emptyset$	$\emptyset$
r_{15}	overlaps	$\neg\emptyset$	$\neg\emptyset$	$\neg\emptyset$	$\neg\emptyset$
r_7	included-at-border	$\neg\emptyset$	$\neg\emptyset$	$\neg\emptyset$	$\emptyset$
r_6	included	$\emptyset$	$\neg\emptyset$	$\neg\emptyset$	$\emptyset$
r_{11}	contains-at-border	$\neg\emptyset$	$\neg\emptyset$	$\emptyset$	$\neg\emptyset$
r_{10}	contains	$\emptyset$	$\neg\emptyset$	$\emptyset$	$\neg\emptyset$
r_3	equal	$\neg\emptyset$	$\neg\emptyset$	$\emptyset$	$\emptyset$

Table 4.2: Specification of binary topological relations

Some other properties are immediately evident. The relations $\{d, t, o, =\}$ are symmetric, e.g., A[d]B iff B[d]A. The containment relations $\{c@b, c, i@b, i\}$ are pairwise converses, e.g., A[c]B iff B[i]A. Finally the containment and equality relations make a statement about relative size and shape, e.g., only a larger object of appropriate shape can contain a smaller one, and only objects of the same size and shape can be equal. (See also Figure 4.10 below.)

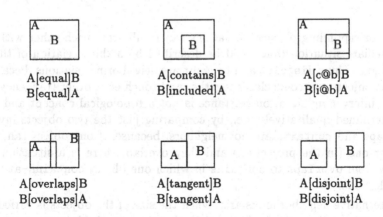

A[equal]B
B[equal]A

A[contains]B
B[included]A

A[c@b]B
B[i@b]A

A[overlaps]B
B[overlaps]A

A[tangent]B
B[tangent]A

A[disjoint]B
B[disjoint]A

Figure 4.9: Topological relations

4.3.3 Structure of the topological domain

The next question to ask is what the neighborhood structure of topological relations looks like, given that we are interested in positional information only. As was pointed out in section 2.3 the structure of the relational domain results in this case from the structure of physical space. The neighborhood shown in Figure 4.10 assumes objects of fixed size that "move around", as seems to be most appropriate for positional information[8] (other neighborhoods are shown below). Two relations are directly linked by an arc in the figure (neighbors)

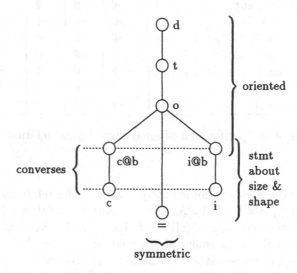

Figure 4.10: Structure of the topological domain

if the corresponding physical situations can result from each other without an intermediate situation that could be described by a third relation of the same dimension. Thus, disjoint and tangent are neighboring relations, because two disjoint objects can come closer together and touch each other (tangency) without an intervening situation (distance is *not* a topological concept and cannot be determined qualitatively, i.e., by comparing just the two objects involved). Overlaps and contains are not neighbors, because, if one object can contain another one (in the projection) and they overlap, there is a situation in the change from overlaps to contains in which one object contains-at-border the other one.

Alternative neighborhoods arise when the size of the objects is varied, while their position is kept fixed. Figure 4.11 shows two such alternative structures, left, the case where only one object size is varied, right, the case where both object sizes are varied. Since we are primarily interested in positional information, these neighborhoods will not be considered further, and are shown only to make the point that more than one neighborhood structure is possible.

Neighboring relations behave similarly (i.e., have similar compositions), which allows us to define hierarchically organized levels of granularity by considering groups of neighboring relations as a single "coarse" relation. These "levels of granularity" are organized hierarchically in homogeneous layers. The topological

---

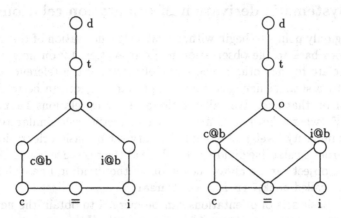

Figure 4.11: Alternative neighborhood structures

levels are organized as follows:

$$
\left.\begin{array}{l}
\left.\begin{array}{l}
\texttt{disjoint} \\
\texttt{tangent}
\end{array}\right\} \texttt{no-contact} \\
\left.\begin{array}{l}
\texttt{overlaps} \\
\texttt{included-at-border} \\
\texttt{included} \\
\texttt{contains-at-border} \\
\texttt{contains} \\
\texttt{equal}
\end{array}\right\} \texttt{contact}
\end{array}\right\} \texttt{no topological info}
$$

Other organizations are possible, for example $\{o,t,d\}$ = non-containment vs. $\{i@b, i, c@b, c, =\}$ = containment or $\{i@b, i, c@b, c, o, =\}$ = projections vs. $\{d, t\}$ = solids.

Topological information alone is insufficient to express positional information, because topology, by its very nature, abstracts from some essential properties such as orientation. In the next section we introduce orientation as the second component in our positional model.

4.4 Orientation

Orientation relations describe where objects are placed relative to one another. Three elements are needed to establish an orientation: a primary object, a reference object and a frame of reference. In the following sections we will derive sets of orientation relations at various granularity levels, and discuss their structure, assuming for simplicity objects with no extension (i.e., points). Section 4.4.3 discusses the three types of reference frames commonly used, and introduces notation for writing down relative positions. Finally, section 4.4.4 presents the mechanisms necessary to handle extended objects.

4.4.1 Systematic derivation of orientation relations

Considering only points to begin with, a systematic derivation of the orientation relations goes back to the observation made in section 4.2 on how 3 points in the plane relate to each other: The point of view and the reference object are connected by a straight line such that the primary object can be to the left, to the right or on that line. We call the three resulting relations $left_1$, $right_1$, $collinear_1^{lr}$ (where the indices are needed to distinguish similar relations at different granularity levels). A complementary set of basic orientations results from the perpendicular line dividing the plane at the reference point. Here again, the primary object can be above, below or on the dividing line. The resulting relations are called $back_1$, $front_1$, $collinear_1^{fb}$.

The two basic sets of orientations can be merged to obtain the next level of finer distinctions (see also Figure 4.15 on page 44). Here again, two sets of relations are possible. One results from the superposition of the basic sets and comprises the relations $left\text{-}back_2$, $left\text{-}front_2$, $right\text{-}front_2$, and $right\text{-}back_2$. Its main application is in "rectangular" domains such as the document layouts studied by Fujihara and Mukerjee (1991) (see section 8.1). The other, more common one, requires in addition to the superposition a rotation of the axes by 45 degrees and contains the relations $back_2$, $left_2$, $front_2$, and $right_2$. Collinearity, which plays a key role on the first level to handle the case in which no distinction can be made between $left_1/right_1$ or $back_1/front_1$ (particularly as the result of composition, see section 5.2), is represented only implicitly at level 2. The assumption is that the pairs of relations $left_2/right_2$ and $back_2/front_2$ complement each other, such that if no distinction is possible between $left_2$ and $right_2$, then it must be $back_2$ or $front_2$, and vice versa.

The next level with the eight distinctions (abbreviations in parentheses) $front_3$ (f_3), $back_3$ (b_3), $left_3$ (l_3), $right_3$ (r_3), $left\text{-}back_3$ (lb_3), $right\text{-}back_3$ (rb_3), $left\text{-}front_3$ (lf_3), and $right\text{-}front_3$ (rf_3) is built in a similar manner by superimposing and rotating the axes from the previous level.[9] Since this third level is the most commonly used, the indices are normally omitted and used only for levels 1 and 2. Figure 4.12 shows examples of objects related by orientations at this level.

Note that in contrast to arrangement, the dividing lines are not dictated by the individual points involved in a particular configuration but result from the projection of distinguished axes onto the reference object. In this anthropocentric orientation system an external observer transfers his characteristic set of axes to the reference object.

At each level, the sets of relations given above are complete and their elements mutually exclusive. All 30 orientation relations $(2 * (3 + 4 + 8))$ are "independent" in the sense that they can be simultaneously members of a disjunction of possible orientations between two objects. Of course, finer relations imply the coarser

---

[9]The case of superposition alone containing the relations $back\text{-}left_3'$, $left\text{-}back_3'$, $left\text{-}front_3'$, $front\text{-}left_3'$, $front\text{-}right_3'$, $right\text{-}front_3'$, $right\text{-}back_3'$, and $back\text{-}right_3'$ (see structure on the left, level 3 in Figure 4.15) has less practical relevance and will not be considered any further in our examples.

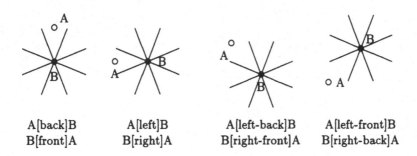

A[back]B   A[left]B   A[left-back]B   A[left-front]B
B[front]A B[right]A B[right-front]A B[right-back]A

Figure 4.12: Orientation relations at level 3

relations with which they share a region (see next section for further details on how orientation at different levels are related). At levels 2 and 3, all relations are assumed to have the same resolution, i.e., **left-back**, for example, is not a finer distinction than say **back**. For our purposes it is not necessary to make finer orientation distinctions, even though some existing qualitative navigational systems have 16 or even 32 different orientations. The formal properties of the domain and the applicability of the approach do not depend on the exact number of orientations. Rather than recurring to finer distinctions, descriptions often change their reference objects to obtain a higher degree of accuracy when necessary. Furthermore, multiple constraints (see section 5.2.5) allow an arbitrarily precise localization of objects without increasing the number of basic relations.

The various levels highlight a further difference between the orientation relations and their linguistic counterparts: In the typical use of spatial prepositions such as "behind" (corresponding to our **back**), the level information is not stated explicitly, and must be derived from the textual or situational context. Thus, if the relations are to represent verbal descriptions, the translation procedure must determine the appropriate level for the corresponding relations.

As an aside, the cognitive importance of these axes varies (see Retz-Schmidt (1988) for pointers to the original literature): The vertical axis (which we are not concerned with in 2-D) has a privileged status, because it is fixed by the gravitation of the earth. The front/back distinction is special, because of the asymmetry of the human body, whereas the left/right distinction is less salient (and bound to be confused more often). In a geocentric orientation system (north/south, east/west and subdivisions) the axes are fixed by the physical poles of the earth defining due north.

It is possible to reconstruct the orientations at any level of granularity using the basic +, −, 0 distinctions by introducing virtual points of view. Together with the reference point, these auxiliary points establish additional axes (Figure 4.13 shows these axes for one set of relations from each level; see also Figure 4.15). The signs with respect to these axes define a given orientation. The tables below each figure show the signs with respect to the corresponding axes. Note that for the third level only two signs (shown shaded) at the time are

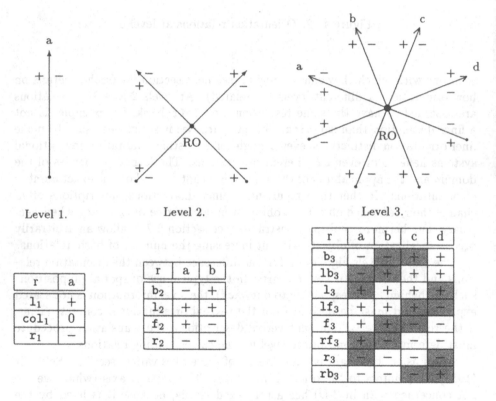

Figure 4.13: Derivation of orientations at different levels

required to fully characterize an orientation: back, for example, is fully specified by $b = -$ and $c = +$, the other signs follow from the ray order a, b, c, d. Neighboring relations differ in one sign, opposite relations in both signs.

Thus the relationship between arrangement and orientation can be summarized as follows:

- Given a simple arrangement relation relating 3 points, an equivalent orientation can be given at the coarsest level.

- Given two position vectors of differing locations in an arbitrary configuration (as defined in section 4.2), both might correspond to the same orientation. With other words, the "resolution" of the orientations is too low to distinguish among locations.

- Given an orientation, the corresponding locations can be constructed by introducing virtual points of view as explained above.

4.4.2   Structure of the orientation domain

Orientations have a uniform circular neighboring structure on each level, as shown in Figure 4.14 for the finest level (as a data structure, this is called a "ron" for "relative orientation node" and will be used in section 5.2.2).[10] Except for

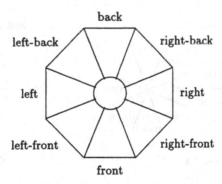

Figure 4.14: Structure of the orientation dimension (ron)

the basic level, each orientation has two neighboring orientations, for example $neigh(b_3)=\{lb_3, rb_3\}$. Except for collinearity, each orientation has an opposite orientation such that if A α B, then B $opp(\alpha)$ A, for example $opp(b_3)=f_3$. On levels 2 and 3, each orientation also has a corresponding set of orthogonal orientations, for example $orth(b_3)=\{l_3, r_3\}$. The "distance" between two orientations in defined as the number of steps necessary to go from one orientation

---

[10] As in the case of topological relations, other types of neighborhood are also possible, where opposite or orthogonal orientations are considered neighbors (as, for example, in some board games). These alternative neighborhoods, however, are not relevant to our discussion, because the positions of physical objects we are interested in change "smoothly".

to the other along the circular structure. On the 3rd level, for example, the distance between an orientation and its opposite orientation is 4, between an orientation and its orthogonal orientation is 2 and so on.

The hierarchical structure follows from the systematic derivation presented in the previous section. However, the relation between orientations at different levels of granularity is not a straightforward one: coarse relations are not just aggregations of finer distinctions as in the topological case. For example, the area denoted by left$_2$ is smaller than the one denoted by the disjunction of lb$_3$, l$_3$ and lf$_3$. As we will see in the next chapter on reasoning, it is often necessary to relate orientations at different levels. A solution proposed by Högg and Schwarzer (1991) introduces a subdivision in 16 sectors, which is fine enough to express any of the relations of the 3 levels considered here as a range of sectors (Figure 4.15). For objects with extension, Kobler (1992) has proposed a 20 sector model (see section 4.4.4). The reasoning mechanisms operate then on the internal range representation and leave it up to context dependent procedures to translate the resulting range into an orientation of an appropriate level.

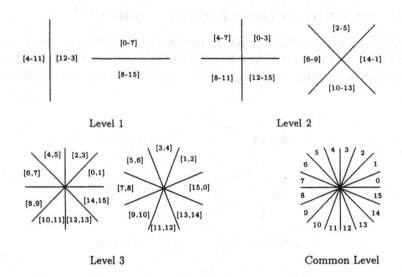

Figure 4.15: Range representation of orientations (adapted from Figure 11 in Högg and Schwarzer 1991)

4.4.3   Reference frames

An important aspect has not been mentioned yet: Orientations must be given with respect to a reference frame, i.e., *"the orientation that determines the direction in which the primary object is located in relation to the reference object"* (Retz-Schmidt 1988, p. 95). Given for example a sentence like *The ball is in front of the car*, studies of the use of projective spatial prepositions in natural language reveal three types of reference frames:

intrinsic: when the orientation is given by some inherent property of the reference object (e.g., *The ball is in front w.r.t. the car front*); Criteria for determining the intrinsic orientation of objects and places are among others: the characteristic direction of motion or use, the side containing perceptual apparatus, the side characteristically oriented towards the observer, the symmetry of objects;

extrinsic: when external factors impose an orientation on the reference object; relevant factors are for example the accessibility of the reference object, its motion (or that of the observer), other objects in its vicinity or the earth gravitation (e.g., *The ball is in front w.r.t. the actual direction of the motion of the car.*, thus if the car is moving backwards, that direction is considered "front");

deictic: when the orientation is imposed by the point of view from which the reference object is seen (from *within* the scene, e.g., by an internal observer or from the speaker's point of view).

Thus, we maintain knowledge about the relative position of two objects that could be declaratively stated as follows:

`<primary_object, [topological,orientation], ref_object, ref_frame>`

where the reference frame with respect to which the orientation is determined can be:

1. *implicit*, i.e., the intrinsic orientation of the parent object (the one that includes both, primary object and reference object) is used as reference. For example, the orientation of the objects in an office is given implicitly by the intrinsic orientation of the office as a whole (which in turn could be determined by such factors as typical entrance or placement of windows).

2. *explicit*, i.e., of the form:

$$\{\text{type-of-use } [, \text{ dex-relation } [,\text{trans}]]\}$$

An explicit reference frame must specify its `type-of-use`, which can be one of intrinsic, extrinsic or deictic as explained above, and optionally (indicated by double square brackets), if the `type-of-use` is not `intrinsic`, the `dex-relation`, which relates the external factor or observer to the reference object. A further `trans`-formation might be required, to determine the `back/front` sides of the reference object with respect to the observer. "Mirror" transformation: fronts point to each other, backs point in opposite directions; "tandem" transformation: both fronts and both backs point in respective common directions.[11]

---

[11]This is mentioned only for the sake of completeness, since most languages hold to the mirror principle, which we will assume as the default from now on.

The implicit orientation is used as the canonical reference frame. That is, for reasoning purposes, all relations between objects are converted first into the implicit form. This is the most natural representation for most applications (see examples in section 4.5). It corresponds, for example, to the act of an external observer of viewing the layout plan of an office from above such that the labels (assumed to convey the orientation of the office as a whole) are "right side up". On "larger scale" spaces the "preferred" orientation might be given by global directions (north, south, east, west).

4.4.4　Objects with extension

The previous subsections assumed objects without extension (i.e., points) for the sake of simplicity. In most situations, however, we have to deal with extended objects, where the size and shape of objects, and the distance between them, influence the orientations that we would accept as valid descriptions of the position of the objects relative to each other. The area in which a particular orientation is accepted as a valid description of the relative position of two objects is called "acceptance area". Consider, for example, the situation depicted in Figure 4.16: The center points of the blocks on the left and those on the right hand of the figure are more or less at the same absolute position. Accordingly, if only center points are used to determine acceptance areas, in both cases B is considered to be $right_2$ of A, even though $back_2$ seems to be the most appropriate orientation to describe the situation on the left, if the relative sizes and shapes of the objects involved are taken into consideration.

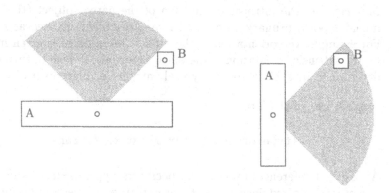

Figure 4.16: Acceptance areas

The distinct roles played by the primary object and the reference object in the localization process also become evident when we consider objects with extension. For example, large and/or salient objects are usually preferred as reference objects as in "The bicycle is in front of the cathedral" as opposed to (∗) "The cathedral is behind the bicycle".

In what follows, we first review previous efforts to handle the orientation of extended objects, and present then our own approach based on heterogeneous non-overlapping acceptance areas.

Previous Work

Early work by Evans (1968) and Haar (1976) handled extended objects by determining the relative orientation of the corresponding centroids. In the example shown in Figure 4.17, P_2 is right of P_1 if the centroid of P_2 falls within the triangular area extending to the right of P_1's centroid. In the triangular model,

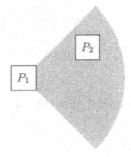

Figure 4.17: Triangular model

as the approach is called, adjoining triangular areas are assumed to have a 50 % overlap to account for the fuzziness of the human judgments that it intends to imitate. The triangular model is a good approximation to handle objects with uniform size and shape and objects which are far away from each other. In both cases, the results obtained in previous sections can be transferred to extended objects by concentrating on the centroids. The approach fails, however, for objects of different sizes in "close proximity". For example in Figure 4.18, P_2 is not considered to be right of P_1, even though that would be the most acceptable interpretation. As said in section 4.3 distance is not a topological concept. Thus proximity is defined here with respect to the relative sizes and the shape of objects involved: An object is in "close proximity" of another one, if its centroid is inside the maximum radius area of the reference object. The maximum radius area is obtained by rotating the reference object (assumed to be the larger one) on its centroid.

For objects in close proximity Peuquet and Ci-Xiang (1987) developed a modification of the triangular model whereby *"the vertex of the triangular area is moved backward or forward relative to the direction at issue so that the ends of the facing side of the frame of the reference polygon touch the boundaries of the projected triangular sector while preserving the original vertex angle"*. Figure 4.19 illustrates this method. In some cases, in which the frame of the primary object is only partially contained in an acceptance area, reversing the

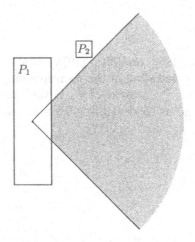

Figure 4.18: Objects of different sizes in "close proximity"

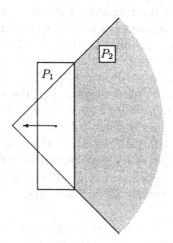

Figure 4.19: Modified triangular model

roles of primary and reference objects and using the opposite orientation might yield an unambiguous description. That description is probably better than the original because such situations arise often when the smaller object is used as reference object contrary to the usual preference. Peuquet and Ci-Xiang (1987) also consider the case of intertwined polygons, i.e., configurations where the centroid of either object falls within the frame of the other object (i.e., the delineative rectangle enclosing it). Their heuristic procedure for that case is very complex and does not handle all cases. For overlapping objects Kobler (1992) proposed an asymmetric subdivision of the acceptance areas with respect to the delineative rectangle containing the reference object. The corners of the rectangle have smaller, the sides have larger corresponding acceptance areas (Figure 4.20). An orientation is established based on the asymmetric sector where the defined center[12] of the primary object falls within.

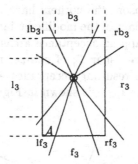

Figure 4.20: Acceptance areas for overlapping objects (adapted from Kobler 1992, p. 33)

Heterogeneous non-overlapping acceptance areas

One difficulty arising from the approaches described above is that the requirement of mutual exclusion of orientations is dropped. Whether we start with overlapping areas as in the original triangular approach or we get them through the translation process used in the modified triangular approach, the fact is, that we might end up with two orientations describing the same situation. Furthermore, the definition of "close proximity" based on the maximum radius area of the reference object yields an area that is too small, particularly for the kind of "well-proportioned" shapes likely to occur in our example domain of office layouts.

The model we propose is a blend of the previous approaches that avoids these difficulties. We distinguish three different cases: objects *far away* from each other, objects in *close proximity*, and *overlapping* objects. A primary object is said to be in "close proximity" of a reference object, if its centroid falls within

---

[12]Kobler (1992) defines the center based on the object's function. This center does not necessarily coincide with the geometrical centroid.

an area up to three times the maximum radius of the reference object. This, of course, is an ad hoc value for which there is no justification other than that it seems to work in the application domain we have been using. Proximity, as a qualitative concept, depends on many factors, some of them not even spatial (cf. section 7.3), which are difficult to account for.

For objects "far away" from each other (i.e., those not in "close proximity" according to the definition above, and, of course, not overlapping), a variant of the triangular model *without* overlapping acceptance areas works fine. That is, the primary object is `left-back` of the reference object if its centroid falls within the triangular area extending in that direction from the centroid of the reference object.

For objects in "close proximity" from each other, the decisive factor is the relative size of the side of the reference object facing the primary object (relative both w.r.t. the other sides of the reference object and the sides of the primary object). For the same reason, the corners have to be treated specially. Both goals are achieved by a variant of the modified triangular model described above, that does not allow overlapping acceptance areas. Depending on the shape of the reference object (where "shape", of course, means the proportions of the delineative rectangle), the resulting acceptance areas have different sizes and forms (i.e., are *heterogeneous*), as is shown in Figure 4.21. As a consequence of

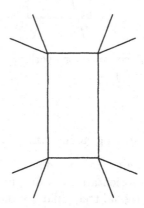

Figure 4.21: Heterogeneous acceptance areas

this method, there is a tendency for elongated objects to allow only orientations at level 2 or even level 1, because of the dominance of one face. In the extreme case of roads or rivers in areal pictures, only $left_1/right_1$ or $back_1/front_1$ are allowed, which corresponds nicely to the usual verbal descriptions of such situations. Finally, for overlapping objects, we adhere to the scheme proposed by Kobler, except that we use the geometric centroid of the reference object as starting point for the acceptance areas and not the functional center point.

To summarize, it can be said that the area of acceptance for a given orientation increases:

- with the distance between primary and reference objects (up to the limit where objects can again be reduced to points);

- with the size of side of reference object facing the primary object.

In cases of ambiguity, that is, in situations where the primary object seems to be on the borderline of two acceptance areas, we recur either to a coarser orientation (which might be a more appropriate description of a situations humans are also likely to describe ambiguously) or to an orientation in the complementary set at the same level. Furthermore, the same mechanisms that help disambiguate the inverse case of one orientation describing different situations (i.e., the standard qualitative "under-determination") can be used here.

4.5 Examples

As an example consider the following natural language descriptions of spatial relations (see Figure 4.22):

(1) Petra's desk (A) is to the left of Daniel's desk (B). (Figure 4.22, view I)

(2) Petra's desk (A) is behind Daniel's desk (B) as seen from Christian's (D) office. (Figure 4.22, view II)

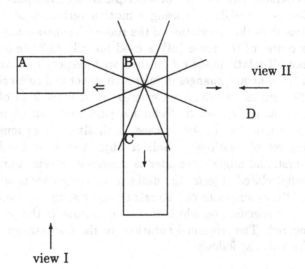

Figure 4.22: Two views on relevant relations

The relative spatial information contained in these sentences can be expressed declaratively as:

(1) REL1 = <A, [d,l], B>

(2) REL1 = <A, [d,b], B, {deictic, REL2}>
    REL2 = <D, [t,b], B>

In (1) no point of view is mentioned, so we assume the scene's implicit orientation which is given by the entrances and the position of the windows (view I in Figure 4.22). To see the difference between implicit and intrinsic orientations, note that A is to the intrinsic *front* of B, since B's intrinsic front as a desk is the side where the drawers open to or where one typically sits (see double arrow in Figure 4.22). However, if we were to store all relations with an intrinsic reference frame, we would have to transform back and forth between the intrinsic and the implicit orientations for most queries. In (2) a point of view is explicitly mentioned. Here an explicit reference frame is necessary to establish the orientation of the point of view with respect to the reference object.

A somehow more involved example is motivated by the following natural language description of an office (see Figure 4.23):

1. *When you enter the room the first thing you see is a large billiard-table (B) on the right, in front of a blackboard (BB).*

2. *On the left, behind a partition (P) there is a group of desks (Ds).*

3. *A copier (C) is next to it.*

Several characteristic features of descriptions are exemplified here. An initial point of view is established by using a motion verb such as "enter", which at the same time fixes the orientation of the room (the entrance is usually the front part). The center of the room (M) is used initially as virtual reference object. However, not all relations are established with respect to that initial reference object. On the contrary, changes in reference object and reference frame are very common. Sometimes, it might not even be clear what kind of reference frame is being used, as in (2), where *"behind a partition"* can be interpreted either implicitly or deictically. In such a case, both alternatives must be included in the resulting set of relations.   Salient objects such as the billiard-table are mentioned first, and might serve later as reference objects. Furthermore, groups of functionally related objects (the desks in this case) are positioned as a single cluster. Finally, descriptions can be rather vague as in *"A copier is next to it."*, where it is not specified on which side of *it* (meant is the *group* of desks) the object is located. The relational notation for these statements with their given reference frames is as follows:

1. <B, [d, r], M>
   <B, [d, f], BB, {intrinsic}>

2. <P, [d, l], M>
   <Ds, [d, b], P>
   <Ds, [d, b], P, {deictic, <M, [d, r], P>}>

3. <C, [{d,t}, {1,r}], Ds>

Where [{d,t},{1,r}] stands for {[d,1],[d,r],[t,1],[t,r]}. In the next chapter (section 5.1), we will discuss the mechanisms required to transform the relations to the canonical frame of reference, resulting in the following set of relations:

1. <B, [d, r], M>
   <B, [d, 1], BB>

2. <P, [d, 1], M>
   <Ds, {[d, b], [d, 1]}, P>

3. <C, [{d,t}, {1,r}], Ds>

It is only with this kind of relational representation that the spatial reasoning mechanism will deal with, and *not* with the complex linguistic issues of translating natural language statements into the relational form.

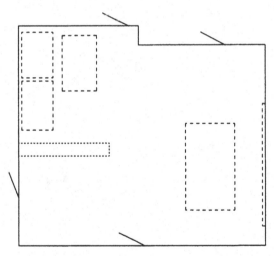

Figure 4.23: A more involved example

4.6 Summary

We focus on 2-D projections of 3-D scenes. In order to represent the relative position of two objects in 2-D space qualitatively we define a small set of spatial relations from the two relevant dimensions topology and orientation. The relative position is given by a topological/orientation relation pair. A complete set of topological relations can be derived from the combinatorial variations of the point set intersection of boundaries and interiors of the involved objects by imposing the constraints of physical space on them. The orientation dimension results from the transfer of distinguished reference axes from an observer to the reference object. Relative orientations must be given w.r.t. a *reference frame*, which can be *intrinsic* (orientation given by some inherent property of the reference object), *extrinsic* (orientation imposed by external factors), or *deictic* (orientation imposed by point of view). When reasoning about orientations, the reference frame is *implicitly* assumed to be the intrinsic orientation of the parent object (i.e., the one containing the objects involved), unless *explicitly* stated otherwise.

Chapter 5

Reasoning with qualitative representations

It is quality rather than quantity that matters.

Lucius Annaeus Seneca, *[Epistles, 1,3], Ib. 45,1.*

The concept of representation introduced in chapter 3 is not a static one. For a representation to be of any use, we have to consider not only its constituents and how they correspond to what is being represented, but also the mechanisms operating on them. In this chapter we look into the mechanisms that allow us to reason with qualitative representations in general, and with the kind of representation of 2-D positions introduced in the previous chapter in particular. These mechanisms are determined in part by the tasks for which qualitative reasoning is used, such as: Inferring knowledge implicit in the knowledge base; answering queries given partial knowledge and a specific context; maintaining various types of consistency; acquiring new knowledge; and, particularly in the case of spatial knowledge, building cognitive maps and visualizing qualitatively represented spatial situations. Some of these tasks will be explored in further detail in chapter 6.

We begin with very simple mechanisms from the spatial domain used to transform between different frames of reference. These transformations are necessary to obtain canonical reference frames, which are a pre-requisite for qualitative inference. We then explore the composition of spatial relations as the basic inference operation. Exploiting the structure of the relational domains makes the regularities in the composition tables evident and suggests efficient ways to compute them. Section 5.3 steps back to introduce very general constraint reasoning mechanisms. The Constraint Satisfaction Problem is identified as an abstract formulation of many difficult problems in AI. An overview of the solution techniques available in the literature is given. Section 5.4 further explores the specific mechanisms required for spatial reasoning. It introduces the concept of abstract maps which allow the solution of some tasks by diagrammatical means. It also shows how, in the particular case of spatial constraints, the structure of the domain greatly reduces the complexity of the algorithms involved. Thus, the general pattern of this chapter is to contrast the reasoning mechanisms known from the literature on relational representations with those that result from exploiting the rich structure of the spatial domain.

5.1   Transforming between frames of reference

Section 4.4.3 in the previous chapter introduced the concept of reference frames and stated that the implicit orientation (i.e., the intrinsic orientation of the parent object) is used as canonical reference frame for reasoning purposes. If a given relation does not have an implicit reference frame, how do we get one? This section explains the mechanisms used to transform the orientations from any other reference frame into an implicit one.

The intrinsic orientation of the spatial entity enclosing the objects being described is the most natural frame of reference for the relative orientations among the objects. In our example domain of 2-D layouts of office spaces, the intrinsic front is generally determined by the main entrance into the room. Other factors determining the intrinsic orientation of a room are, for example, the availability and location of windows, the location of furniture as a whole (particularly if it forms a functional unit), and the location of the room itself relative to other adjacent spaces (particularly if the room is small compared to those neighboring spaces). An absolute orientation of the room with respect to the four cardinal points is almost never used as a frame of reference for the orientations of objects in the room.[1]

Even though an implicit frame of reference is preferred in most typical descriptions, other reference frames are used as well. If, for example, the reference object has a prominent intrinsic orientation, an intrinsic frame of reference is likely to be used. If the reference object does not have an intrinsic orientation at all (e.g., a round table or a tree), an extrinsic reference frame has to be used. A deictic reference frame is sometimes preferred, if the point of view of the speaker embedded in the scene is the focus of attention. In natural language descriptions there is even a frequent change of both reference objects and reference frames. To reason efficiently with spatial relations, however, they have to be converted first into an implicit form. To do so, we need the intrinsic orientation of the parent object and the orientation of the primary object with respect to the reference object in the given frame of reference. The following special cases have to be considered, depending on the type of reference frame (Figure 5.1 shows a summary of the transformation mechanisms):

- **intrinsic** $\longrightarrow$ **implicit:**

---

[1]There is, however, something like an conventional view of maps, whereby north is assumed to be up.

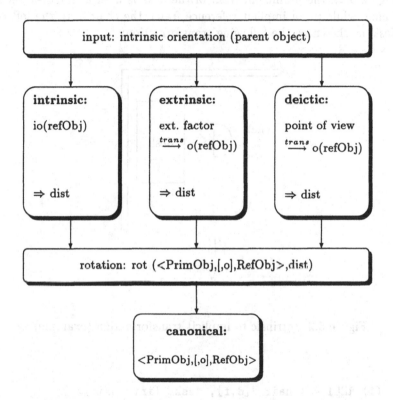

Figure 5.1: Transforming between reference frames

We need the intrinsic orientation of the parent object and the orientation of the primary object w.r.t. the reference object. The number of orientation sectors between those two corresponds to the "distance" that the original relation needs to be rotated to obtain an orientation in the canonical frame of reference.

If the reference object has an intrinsic front and the orientation of the primary object is given with respect to it, then all we need is the relative direction of that intrinsic front w.r.t. the intrinsic orientation of the parent object (i.e., the *distance d* between those two orientations). Given the circular structure of orientations, the implicit orientation of the primary object is obtained by *rotating* the original orientation by the same distance *d*. For example, in the sentence *"The chair is in front of the desk."* (see Figure 5.2), the prominent front of the desk is used as intrinsic reference frame, while in an implicit reference frame the chair is to the left of the desk, as shown by the following transformation:

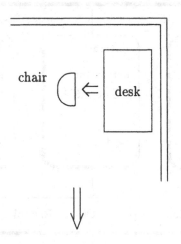

Figure 5.2: Intrinsic to implicit transformation (example)

```
(1) REL1 = <chair, [d,f], desk, {intrinsic}>
    dist (io(office), io(desk)) = -2
    rot (f, -2) = l
    ==> <chair, [d,l], desk>
```

- **extrinsic $\longrightarrow$ implicit:**
  If an external factor imposes an orientation on the reference object, then we need to establish the rotational "distance" between that imposed front and the intrinsic front of the parent object. Once this has been done, the same type of rotation done in the previous case yields the implicit orientation. A typical example of an extrinsic reference frame is the sentence *"The car is in front of the oak tree"*. Since trees do not have an intrinsic front, the reference frame is determined by the situational context. In this case, the tree is in the immediate vicinity of a house, which has a prominent intrinsic front (see Figure 5.3). That orientation is thus transferred to the tree as

reference object. Assuming, for example, that the block in which the house is located is the "parent" object and that it has the intrinsic orientation shown in the figure (fixed, for example, by its geographical orientation), the following transformation leads to the canonical form:

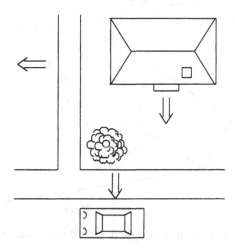

Figure 5.3: Extrinsic to implicit transformation (example)

```
<car, [d, f], tree, {extrinsic, transfer(tree, house)}>
dist (io(block), o(tree)) = +2
rot (f, +2) = r
==> <car, [d, r], tree>
```

- **deictic ⟶ implicit:**
  If the point of view from which the reference object is seen determines its orientation, we have a deictic reference frame. Deictic reference frames are a very frequent special case of extrinsic frames of reference. Thus the same procedure as in the previous case applies. Figure 5.4 shows an example in which in the sentence *"The bicycle is in front of the church"* a reference object with a prominent intrinsic front (the church) is nevertheless used deictically from the point of view of an observer embedded in the scene. Note that the axes induced on the reference object do not necessarily map on its intrinsic axes. Here the intrinsic front of the church is used as implicit orientation as well, since there is no obvious parent object.

```
REL1:<bicycle, [t, f], church, {deictic, REL2}>
REL2:<observer, [d, lb], church>
dist (io(church), o(observer)) = -3
rot (f, +3) = lb
==> <bicycle, [d,lb], church>
```

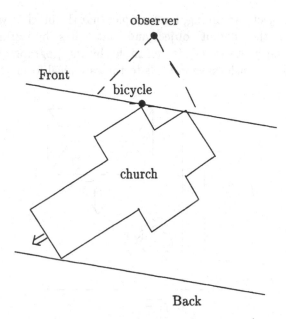

Figure 5.4: Deictic to implicit transformation (example)

Although we have been using sentences to illustrate our examples, the linguistic mechanisms needed to handle the use of reference frames in natural language are much more involved than what the simple transformations presented here suggest. There are competing criteria (e.g., prominent intrinsic orientation vs. speaker and/or hearer deixis) and a strong context dependance in the determination of the proper reference frame (see, e.g., Levelt 1986; André, Bosch, Herzog, and Rist 1987; Retz-Schmidt 1988; Pribbenow 1991, for surveys on the subject). For a study of perspective-taking behavior in spatial mental models acquired from texts see Franklin, Tversky, and Coon (1992).

Several ways of implementing the transformations come to mind. Table lookup is straightforward but a waste of space given the uniform circular structure of orientations. Mapping both the intrinsic orientation of the parent object and the orientation of the reference object to an absolute numerical system allows to "compute" the distance by subtracting the corresponding indices (modulo the number of orientations at the particular granularity level). An equivalent but conceptually more attractive option is to use a data structure with the same circular structure as the orientation domain. The transformation is then achieved by "rotating" the labels by the necessary distance. The advantages of such "analogical" representations become evident when several relations are simultaneously affected by a transformation as in the case of a change in point of view (see section 5.4.2).

5.2 Composition of spatial relations

An essential operation in every relational representation is the composition of relations: Given the relation between A and B, on the one hand, and between B and C, on the other, we want to know the relation between A and C. We will first consider topological relations and orientations separately. In both cases, we will first look at simple composition tables and discuss then how the regularities found there lead to efficient ways to compute them. The composition of the separate domains turns out to be strongly under-determinate, i.e., the resulting relation sets tend to contain too many alternative relations. Fortunately, the composition of positional information, i.e., of topological/orientation pairs, yields more specific results. Furthermore, the effect of multiple, mutually constraining compositions can lead to arbitrarily precise descriptions.

5.2.1 Composition of binary topological relations

The composition of binary topological relations can be derived in a straight-forward manner from the intuitive understanding of their definitions. A formal derivation based on the set theoretical definitions given in section 4.3.1 has been done by Egenhofer (1991). The idea is to reduce the composition of topological relations to the transitivity property of the subset relation. This property is expressed canonically in terms of the set intersections used to define the topological relations, i.e., the intersection $I_z[P_*^A, P_*^C]$ is derived from the intersections $I_x[P_*^A, P_*^B]$ and $I_y[P_*^B, P_*^C]$ defining the corresponding relations $A[R_x]B$ and $B[R_y]C$ (where $P_i^A, P_j^A \in \{\partial A, A^\circ\}$ and $P_i^A \neq P_j^A$; $P_l^B, P_m^B \in \{\partial B, B^\circ\}$ and $P_l^B \neq P_m^B$; $P_o^C, P_p^C \in \{\partial C, C^\circ\}$ and $P_o^C \neq P_p^C$).

Table 5.1 gives an overview of the inference rules for intersections, simplified to include only those intersections needed, given the restricted types of objects considered (see section 4.1).[2] Once the derived intersections have been combined using logical operators, they have to be compared with the eight possible intersections to recover the corresponding topological relations. Table 5.2 shows the result of this lengthy process, for which Egenhofer (1991) used a PROLOG program, adapted to our notation.[3] While correct, this table allows only to recognize a couple of statistical properties of the resulting compositions, such as that only twelve compositions are unique (besides trivial compositions with equal), three contain all relations (d/d, i/c, o/o), two are commutative (i⊛b/i, c⊛b/c), and three relations are transitive (=, i, c). Reorganizing the table according to a useful linearization of the domain structure (as introduced in section 4.3.3) reveals other interesting regularities (see Table 5.3):

---

[2]Egenhofer (1991) uses an extended '9-intersection' model, which he claims to be superior to the '4-intersection' model, because it also relates objects to the embedding space. He then goes on to exclude objects with holes, co-dimension greater than 0 or separations. However, those would be the only cases in which the 9-intersection model would make a difference. For the eight topological relations 9-intersections do not discriminate any further than the 4-intersections, they just make the terms larger! (cf. Mark and Egenhofer 1992)

[3]For the complexities of computing composition tables see also Randell, Cohn, and Cui (1992a).

Transitivity rule	Canonical representation
$A \subseteq B \wedge B \subseteq C \rightarrow A \subseteq C$	$I_x[P_i^A, P_l^B] = \neg\emptyset \;\wedge\; I_x[P_i^A, P_m^B] = \emptyset$ $\wedge\; I_y[P_l^B, P_o^C] = \neg\emptyset \;\wedge\; I_y[P_l^B, P_p^C] = \emptyset$ $\rightarrow\; I_z[P_i^A, P_o^C] = \neg\emptyset \;\wedge\; I_z[P_i^A, P_p^C] = \emptyset$
$A \supseteq B \wedge B \supseteq C \rightarrow A \supseteq C$	$I_x[P_i^A, P_l^B] = \neg\emptyset \;\wedge\; I_x[P_j^A, P_l^B] = \emptyset$ $\wedge\; I_y[P_l^B, P_o^C] = \neg\emptyset \;\wedge\; I_y[P_m^B, P_o^C] = \emptyset$ $\rightarrow\; I_z[P_i^A, P_o^C] = \neg\emptyset \;\wedge\; I_z[P_j^A, P_o^C] = \emptyset$
$A \cap B = \neg\emptyset \wedge B \subseteq C \rightarrow$ $A \cap C = \neg\emptyset$	$I_x[P_i^A, P_l^B] = \neg\emptyset$ $\wedge\; I_y[P_l^B, P_o^C] = \neg\emptyset \;\wedge\; I_y[P_l^B, P_p^C] = \emptyset$ $\rightarrow\; I_z[P_i^A, P_o^C] = \neg\emptyset$
$A \supseteq B \wedge B \cap C = \neg\emptyset \rightarrow$ $A \cap C = \neg\emptyset$	$I_x[P_i^A, P_l^B] = \neg\emptyset \;\wedge\; I_x[P_j^A, P_l^B] = \emptyset$ $\wedge\; I_y[P_l^B, P_o^C] = \neg\emptyset$ $\rightarrow\; I_z[P_i^A, P_o^C] = \neg\emptyset$
$A \cap B = \neg\emptyset \wedge B \sqsubseteq (C_0 \cup C_1) \rightarrow$ $A \cap (C_0 \cup C_1) = \neg\emptyset$	$I_x[P_i^A, P_l^B] = \neg\emptyset$ $\wedge\quad I_y[P_l^B, P_o^C] = \neg\emptyset \;\wedge\; I_y[P_l^B, P_p^C] = \neg\emptyset$ $\rightarrow\; \neg(I_z[P_i^A, P_o^C] = \emptyset \;\wedge\; I_z[P_i^A, P_p^C] = \emptyset)$
$(A_0 \cup A_1) \sqsupseteq B \wedge B \cap C = \neg\emptyset \rightarrow$ $(A_0 \cup A_1) \cap C = \neg\emptyset$	$I_x[P_i^A, P_l^B] = \neg\emptyset \;\wedge\; I_x[P_j^A, P_l^B] = \neg\emptyset$ $\wedge\quad I_y[P_l^B, P_o^C] = \neg\emptyset$ $\rightarrow\; \neg(I_z[P_o^C, P_i^A] = \emptyset \;\wedge\; I_z[P_o^C, P_j^A] = \emptyset)$

Table 5.1: Inference rules for intersections

$\sqsubseteq$ is the subset relation between a set A and the union of the sets B and C such that $A \subseteq (B \cup C)$ and $A \cap B = \neg\emptyset$ and $A \cap C = \neg\emptyset$. $\sqsupseteq$ is defined accordingly (adapted from Egenhofer 1991).

	d	t	=	i	i⊕b	c	c⊕b	o
d	d V t V = V i V i⊕b V c V c⊕b V o	d V t V i V i⊕b V o	d	d V t V i V i⊕b V o	d V t V i V i⊕b V o	d	d	d V t V i V i⊕b V o
t	d V t V c V c⊕b V o	d V t V = V i⊕b V c⊕b V o	t	i V i⊕b V o	t V i V i⊕b V o	d	d V t	d V t V i V i⊕b V o
=	d	t	=	i	i⊕b	c	c⊕b	o
i	d	d	i	i	i	d V t V = V i V i⊕b V c V c⊕b V o	d V t V i V i⊕b V o	d V t V i V i⊕b V o
i⊕b	d	d V t	i⊕b	i	i V i⊕b	d V t V c V c⊕b V o	d V t V = V i⊕b V c⊕b V o	d V t V i V i⊕b V o
c	d V t V c V c⊕b V o	c V c⊕b V o	c	= V i V i⊕b V c V c⊕b V o	c V c⊕b V o	c	c	c V c⊕b V o
c⊕b	d V t V c V c⊕b V o	t V c V c⊕b V o	c⊕b	i V i⊕b V o	= V i⊕b V c V o	c	c V c⊕b	c V c⊕b V o
o	d V t V c V c⊕b V o	d V t V c V c⊕b V o	o	i V i⊕b V o	i V i⊕b V o	d V t V c V c⊕b V o	d V t V c V c⊕b V o	d V t V = V i V i⊕b V c V c⊕b V o

Table 5.2: Composition of binary topological relations (adapted from Table 2 in Egenhofer 1991)

- As in the temporal case analyzed by Freksa (1992a), all resulting disjunctions of relations are connected (neighbors), i.e., similar techniques can be applied here as well (see sections 5.4.5 and 8.1).

- The table is "almost symmetric", i.e., given any pair of relations A [x,] B and B [y,] C such that not both x and y are elements of {i,i@b,c,c@b}, their composition A [t,] C is the same as the composition of A [y,] B and B [x,] C, except that if an i or i@b occurs in any of x, y, t it must be substituted by c or c@b, respectively and conversely c by i and c@b by i@b. Example: A [o,] B, B [c,] C → A [{d,t,o,c@b,c},] C, so A [i,] B, B [o,] C → A [{d,t,o,i@b,i},] C. (For this reason and to improve the readability of the table, the vertical order of {i, i@b} and {c, c@b} is flipped with respect to the horizontal order.)

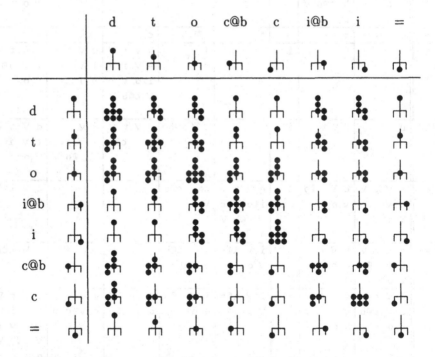

Table 5.3: Composition of binary topological relations arranged according to a useful linearization of the domain structure (the iconic notation is derived from the structure of the topological dimension in Figure 4.10, where ⊥={o, i@b, i}, for example; Table A.1 is the equivalent table using set notation).

Using these regularities, the tables can be coded more efficiently. Furthermore, they provide a primitive form of error correcting consistency check: an error on an early draft of Egenhofer (1991) was immediately apparent after reorganizing the table, because it would have been the only unconnected composition in the table.

Interestingly, not all $2^8 = 256$ possible variations of relations (not even all of the 64 members of the subset of connected relations sets) actually occur as compositions. Only 21 different relation sets result from the first composition step, including the original eight single relations. A "fixed-point" is reached after two iterations and contains a closed set of only 26 different sets of relations, all of them connected (see Table A.2).

5.2.2 Composition of orientation relations

A□B	B□C							
	b	lb	l	lf	f	rf	r	rb
b	{b}	{b,lb}	{b,lb,l}	{b,lb,l,lf}	{?}	{b,rb,r,rf}	{b,rb,r}	{b,rb}
lb	{lb,b}	{lb}	{lb,l}	{lb,l,lf}	{lb,l,lf,f}	{?}	{lb,b,rb,r}	{lb,b,rb}
l	{l,lb,b}	{l,lb}	{l}	{l,lf}	{l,lf,f}	{l,lf,f,rf}	{?}	{l,lb,b,rb}
lf	{lf,l,lb,b}	{lf,l,lb}	{lf,l}	{lf}	{lf,f}	{lf,f,rf}	{lf,f,rf,r}	{?}
f	{?}	{f,lf,l,lb}	{f,lf,l}	{f,lf}	{f}	{f,rf}	{f,rf,r}	{f,rf,r,rb}
rf	{rf,r,rb,b}	{?}	{rf,f,lf,l}	{rf,f,lf}	{rf,f}	{rf}	{rf,r}	{rf,r,rb}
r	{r,rb,b}	{r,rb,b,lb}	{?}	{r,rf,f,lf}	{r,rf,f}	{r,rf}	{r}	{r,rb}
rb	{rb,b}	{rb,b,lb}	{rb,b,lb,l}	{?}	{rb,r,rf,f}	{rb,r,rf}	{rb,r}	{rb}

Table 5.4: Composition of orientation relations

Table 5.4 contains the compositions of orientation relations (recall from section 4.1.2 that, when considered independently, orientation relations ignore extension, i.e., objects are assumed to be ideal points) for the finest granularity level currently used. The notation {l,lb,b} means that any of **left**, **left-back** or **back** might be the case. The regularities here are straightforward, and can be nicely computed diagrammatically as follows: Given A $[,x]$ B and B $[,y]$ C, the composition A $[,t]$ C contains all orientations that are "in-between" and including x and y on the "shortest path" of the "ron" (see Figure 4.14, p. 43). The composition is symmetric, i.e., $t(x,y)=t(y,x)$. The following examples illustrate this rule:

1. If $x = y$ then $t = x = y$, since there are no orientations in-between ("distance" between orientations $= 0$).

2. If A is to the **back** of B and B is to the **left** of C, A can be to the **back**, **left-back** or **left** of C as the three square configurations on Figure 5.5 show.

3. For $x = $ **lb** and $y = $ **r** we obtain $t = $ {lb,b,rb,r} (*not* {lb,l,lf,f,rf,r} which are the orientations between x and y on the longer path "the other way around"). Here, $0 < $ distance $ < 4$.

4. If x and y are opposites (e.g., b/f, l/r, lb/rf) then $t = \{?\}$, that is *any* of
the orientations apply, since both paths between opposites have the same
length (distance $= 4$).

Again, the relations in the composition are always "conceptual neighbors".

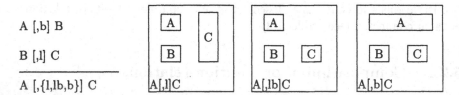

A [,b] B

B [,l] C
_____

A [,{l,lb,b}] C

A[,l]C   A[,lb]C   A[,b]C

Figure 5.5: Composition of orientation relations

5.2.3   Composition of topological/orientation pairs

As can be seen from the two previous sections, the composition of the separate domains yields in part highly unconstrained results, i.e., sets containing too many alternative relations. One might expect this situation to get worse for the composition of topological/orientation pairs. Surprisingly, this is not the case. On the contrary, the resulting sets contain fewer relations. On second thought, it is obvious that positional information is more constrained than topological or orientation information alone. This situation is similar to the one discovered by Waltz (1975) for the computer interpretation of line drawings of blocks: Introducing shadows, a step that seems to make the task more difficult, actually makes it easier, because shadows provide additional constraints equivalent to those resulting from a second viewpoint.

We give here two examples of the constraining effect of combining the two dimensions (further examples can be found in the composition tables in appendix A).

Table 5.5 shows the composition of binary topological relations for an assumed constant orientation. A comparison with Table 5.3 indicates differences for relation pairs having any of $\{d,t,o,i@b,c@b\}$ as their members. In particular the composition of relation pairs involving $\{d,t,o\}$ in both arguments, which is strongly under-determinate in the topological case, yields here sparse or even unique results. Of course, the unoriented containment relations $\{i,c,=\}$ are not affected by the combination with orientations. Relation pairs containing $\{i@b,c@b\}$ yield compositions without containment relations (not even "at border"), except when compared with themselves. The regularities of the combined table are the same as in the case of topological relations alone. There is the same kind of substitution symmetry, provided that not both argument relations are elements of $\{i,i@b,c,c@b\}$.

For example, given the topological relations A [o,] B and B [o,] C (the orientation, say back, is assumed constant for both given relations *and* for the re-

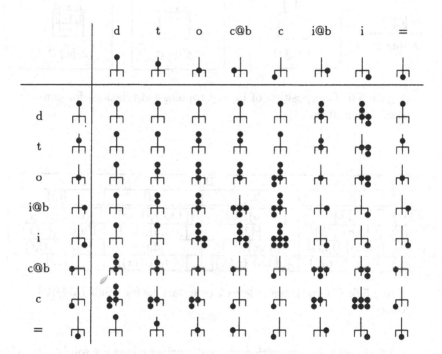

Table 5.5: Composition of topological relations for constant orientation. For example, if A is 'disjoint and in front of' B (A [d,f] B) and B is 'disjoint and in front of' C (B [d,f] C), then A is 'in front of' C (by constant orientation assumption) and 'disjoint of' C (because the non-zero extension of B lying between A and C guarantees that the extensions of A and C are separate).

sult!), which can correspond to any of the three (qualitatively) different square configurations shown in Figure 5.6, the resulting relation between A and C is A [{d,t,o},] C, as can be confirmed by looking at the figures.

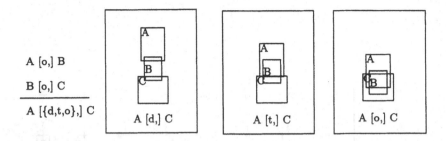

A [o,] B

B [o,] C

A [{d,t,o},] C

A [d,] C   A [t,] C   A [o,] C

Figure 5.6: Composition of binary topological relations for constant orientation

	B∥C			
	[t,b]	[t,l]	[t,f]	[t,r]
[d,b]	[d,b]	[{d,t,o},{b,l}]	[{d,t,o},{l,b,r}]	[{d,t,o},{b,r}]
[d,l]	[{d,t,o},{l,b}]	[d,l]	[{d,t,o},{l,f}]	[{d,t,o},{l,f,b}]
[d,f]	[{d,t,o},{l,f,r}]	[{d,t,o},{l,f}]	[d,f]	[{d,t,o},{f,r}]
[d,r]	[{d,t,o},{r,b}]	[{d,t,o},{r,f,b}]	[{d,t,o},{f,r}]	[d,r]

Table 5.6: Composition of level 2 orientations for solids ([d,]/[t,] case)

Table 5.6 shows the composition of level 2 orientations for solids, i.e., where disjoint and tangent are the only two topological relations allowed (in the argument). The table shows only the quadrant corresponding to the composition of pairs containing d and t respectively, because of the high similarity of the resulting compositions in the four quadrants t/t, t/d, d/t and d/d. Each quadrant itself is pseudo-symmetric on its diagonal. Disjoint and tangent being two of the less informative topological relations, as far as position is concerned (cf. section 5.4.3), the constraining effect of the combined composition is nevertheless clear: Instead of {d,t,o,i⊘b,i}, which we would obtain without orientation, the topological part is restricted to {d,t,o} and in some cases even to d. Likewise, orientation compositions that would yield no information of their own such as f/b and l/r, yield {l,f,r} and {f,l,b} respectively.

5.2.4 Structure and table lookup

One possible objection concerning the mechanisms for the composition of relations is that composition tables grow as the square function of the number of relations involved. This is true, but not a serious drawback in our case, because the number of relations is small. Even when computing the fixed-points of the composition tables, the highest number of relations involved is 26 for topological relations and 33 for orientation relations (level 3). Not all of the relations of the full closed composition set make sense in the combined topological/orientation tables because of the mutual constraints of positional information. Furthermore, all the composition tables have multiple symmetries, which allow a large reduction in the amount of memory required to store them. Freksa (1992a) demonstrates this impressively for the case of temporal relations among semi-intervals.

Another type of objection goes in the opposite direction and criticizes the intensive use of structure, either in form of a compactification of the composition tables or in form of analogical data structures, arguing that there is nothing more efficient than table lookup and that memory is cheap. While this too is true for single compositions, ignoring the structure of relational domain leads to increased computation in cases of ambiguity or uncertainty, because all combinatoric possibilities have to be checked instead of only those that make sense.

5.2.5   The effect of multiple constraints

Despite the constraining effects of the combined composition of topological/-orientation pairs, computing their composition repeatedly seems to lead to a serious "information degradation" in the sense that the resulting relations are increasingly unspecific. Note, however, that compositions are not computed in isolation from other relations possibly constraining them. Consider, for example, the situation depicted in Figure 5.7a. Given the relation between C and D as C [d,1f] D, the fact that E is to the back of C (E [d,b] C) constrains the relation between E and D only to be "somewhere to the left or to the back" (E [d,{1,1b,b}] D). But if we additionally know that E is to the left of A (E [d,1] A), which in turn is to the back of D (A [d,b] D)—see Figure 5.7b—then E is further constrained to be "left-back" with respect to D (E [d,1b] D).

In general the effects of multiple constraints can lead to an arbitrarily precise description of the relative positions of objects. Of course, the general constraint satisfaction problem is known to be computationally intractable. That is, the amount of computation needed to establish consistency in a network of relations grows exponentially with the number of relations. In the next section, we will review general methods to limit the number of irrelevant cases considered.

5.3   Constraint reasoning

Given that qualitative representations can be expressed in form of relations, general constraint satisfaction techniques can be applied for various kinds of inference. Constraint satisfaction techniques play an important role in Computer Science as a whole, an in AI in particular. Many difficult problems involving search from areas such as machine vision, temporal reasoning, scheduling, graph

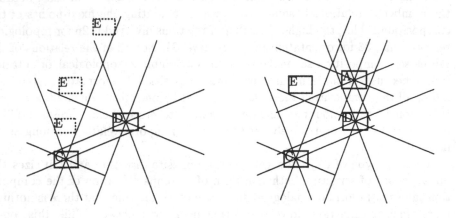

(a) E[]D ambiguous given C[]D, C[]E   (b) Adding E[]A further constrains E[]D

Figure 5.7: The effect of multiple constraints

algorithms, and machine design and manufacturing can be considered to be special cases of the constraint satisfaction problem (CSP).

In this section we define the CSP formally and provide an overview of the solution techniques available in the literature. This overview borrows heavily from excellent review articles by Mackworth (1987), Meseguer (1989), and Kumar (1992). For the special case of geometric constraints see Kramer (1990) and du Verdier (1993). Even though the solution techniques described represent large efficiency improvements over the obvious backtracking algorithm, they are "limited by their generality". That is, being general, domain-independent techniques, they ignore the structure of the relational domain. In section 5.4 we show that taking the structure of the richly constrained spatial domain into consideration leads to more efficient algorithms.

5.3.1 Constraint satisfaction problem

The *constraint satisfaction problem* can be formulated in very general terms as follows:

Given are a set of variables $\{X_1, \ldots, X_n\}$, a discrete and finite domain for each variable $\{D_1, \ldots, D_n\}$, and a set $\{R_k\}$ of constraints, defined over some subset of the variable domains, $R_j \subseteq D_{i1} \times \cdots \times D_{ij}$, and showing the mutually compatible values for a variable subset $\{X_{i1}, \ldots, X_{ij}\}$. The problem is to find an assignment of values to variables such that all constraints are satisfied. (Variants are to find all such assignments, the best one, if there exists any at all, etc.) Without lost of generality we restrict constraints to be unary or binary (Nudel 1983) and assume that all variable domains have the same cardinality.

A labeled directed graph (Figure 5.8) can be used to depict an instance of a CSP by making nodes correspond to variables and arcs to binary constraints.

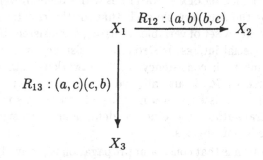

Figure 5.8: A CSP graph

The straightforward approach to find a satisfying assignment (besides the brute-force "generate-and-test" approach) is a backtracking algorithm that corresponds to an uninformed systematic search. If after having instantiated all variables relevant to a set of constraints any of them is not valid, the algorithm backtracks to the most recently instantiated variable that still has untried values available. The run-time complexity of this algorithm is exponential,[4] making it useless for realistic input sizes. This inefficiency arises because the same computations are repeated unnecessarily many times.

The following sections show ways in which the efficiency of constraint satisfaction algorithms can be improved. These improved algorithms can be roughly classified in those that modify the search space and those that use heuristics to guide the search. Two further ways of improvement are parallelization and the use of the particular problem structure to orient the search. While we won't say anything about the former, the latter will be discussed at length in section 5.4.

5.3.2 Consistency improvement

The goal of modifying the search space is to avoid useless computation without missing any of the solutions of the original space. In other words, we are looking for a smaller but equivalent search space. This can be achieved either prior to the search (by improving the consistency of the network through constraint propagation) or, in hybrid algorithms, during the search.

Constraint propagation

One way of reducing the number of repeated computations is to remove those values from a domain that do not satisfy the corresponding unary predicates, as well as those values for which no matching value can be found in the adjacent domains such that the corresponding binary predicates are satisfied. The former

---

[4]$O(a^n)$ to be more precise, where a = cardinality of the domain and n = number of variables.

process achieves *node consistency*, the latter *arc consistency*. This concept of local consistency can be generalized to any number of variables. A set of variables is *k-consistent* if for each set of $k-1$ variables with satisfying values, it is possible to find a value for the kth variable such that all the constraints among the k variables are satisfied. A set of variables is strong k-consistent if it is j-consistent for all $j \leq k$. Of special interest is strong 3-consistency, which is equivalent to arc consistency plus path consistency.   A network is *path consistent* if any value pair permitted by R_{ij} is also allowed by any other path from i to j. The process of achieving consistency in a network of constraints is called *constraint propagation*. Several authors give different definitions of constraint propagation, whose equivalence is not obvious.

One such definition is that constraint propagation is a way of deriving stronger (i.e., more restrictive) constraints by analyzing sets of variables and their related constraints. The value elimination consistency procedures mentioned above assume a very general extensional form of constraints as sets of satisfying value pairs. Whenever a value of the domain of a variable involved in more than one constraint is removed, a satisfying value pair might have to be removed from one or more of the other constraints, thus making them more restrictive. In our example, the original constraint R_{12} allowed the value pairs $\{(a,b), (b,c)\}$ between X_1 and X_2. Making the domain of X_1 consistent with $R_{13} = \{(a,c),(c,b)\}$ eliminates (b,c) from R_{12} because according to R_{13} the value of X_1 cannot be b. Similarly, making the domain of X_1 consistent with R_{12} eliminates (c,b) from R_{13}. Both constraints allow now fewer value pairs and are thus stronger.

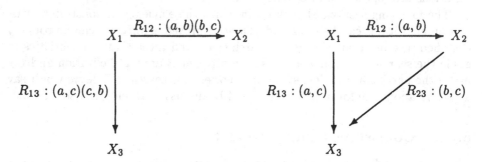

Figure 5.9: Constraint Propagation

Another view of constraint propagation is as the process of making implicit constraints explicit (Montanari 1974; Freuder 1978), where implicit constraints are those not recorded directly in compatible value pairs, but implied by them. This, however, can be seen as a side effect of the consistency procedures. The universal constraint, that allows any value pair, holds implicitly between two variables not explicitly linked together. If the domains of the two variables X_i and X_j are restricted to a few values, then the implicit constraint R_{ij} can be made explicit as the set of combinations of the two domains. In our example, if the domains of X_2 and X_3 have been restricted to {b} and {c} respectively, then the constraint R_{23} can be stated as $\{(b,c)\}$.[5]

Note that a global full constraint propagation is equivalent to finding the minimal graph of the CSP, where each value permitted by any explicit constraint belongs to at least one problem solution. Full constraint propagation is thus as hard as the CSP itself (NP-complete). The local constraint propagation techniques used to achieve node-, arc- and path-consistency have polynomial complexities and can be used as pre-processors that substantially reduce the need to backtrack during the search for a global solution.

Unfortunately, arc- and path-consistency do not eliminate in general the need to backtrack during the search because constraints are propagated only locally. There are, however, some special cases for which backtrack-free search can be obtained. If a total order imposing a temporal sequence of assignments of values to variables is given, then the width of a node is defined as the number of arcs linking it to previous nodes. The width of a graph is the minimum of the widths of all possible node orderings, which in turn are the maxima of the widths of the nodes. If the level of strong consistency of a graph is greater than its width, then there is a backtrack free search order (Freuder 1982). Even though it is relatively easy to determine a minimum width ordering, making that graph strongly $(w + 1)$-consistent adds new arcs (increasing the width of the graph!) for $k > 2$. Thus, except for tree-structured constraint graphs, which have width 1, further improvements are required to obtain reasonably efficient algorithms. Among others, eliminating redundant constraints and using adaptive consistency (requiring different consistency levels for each node depending on its width) have been proposed in the literature (Dechter and Dechter 1987; Dechter and Pearl 1988, respectively).

Hybrid algorithms

The consistency improvement methods described in the previous section modify the search space prior to the actual search. Another class of algorithms combines the search space reduction with the actual search. This can be done in several ways including look-ahead and look-back procedures, graph theoretical methods and dynamic programming techniques.

The basic look-ahead scheme is to use a limited constraint propagation to obtain a desired level of consistency, whenever a new node of the search space is visited. Depending on the consistency level called for, we obtain various degrees of look-ahead behavior ranging from simple backtracking, through forward checking, partial look-ahead and full look-ahead to "really full look-ahead". Interestingly enough, empirical studies have shown that the overhead of achieving higher consistency levels is larger than the benefits obtained. For the well-known

---

[5] A further use of the term constraint propagation can be found in (Steele 1980), where a constraint language is defined. Constraints are viewed there as "physical devices" enforcing simultaneously the relations they represent on the variables "connected" to them (typical examples are arithmetic constraints such as adder, multiplier, etc., i.e., not binary constraints). Actual computation is done by a step-by-step propagation of values from one device (=constraint) to the next, that is somewhat misleadingly called constraint propagation.

n-queens problem, for example, forward checking provides the best results, even though it does not test consistency among future variables (Nudel 1988).

The common idea behind look-back algorithms is to improve backtracking by making it more "intelligent". Ideally backtracking should be done directly to the variable responsible for the inconsistency, while avoiding having to perform redundant work (i.e., rediscovering incompatible values). Dependency-directed backtracking (Stallman and Sussman 1977) does exactly that by recording "justifications" for value assignments. Initially the justifications can be as simple as that "no justification exist for alternative values". Whenever a constraint violation is detected, however, a new node is created denoting the incompatibility of the involved values. This node is then used to justify an alternative value assignment. Again, a full blown implementation of this technique, while reducing search to a minimum, carries an overhead that in many cases makes it slower than simple backtracking. Thus, simpler variants such as backjumping, backmarking and conflict set lists have been developed (Dechter 1986, 1990).

Several algorithms use special properties of the graph representing the constraint network to make the search process more efficient.

5.3.3   Heuristics

The basic backtracking search strategy does not specify the order in which variables are selected for backtracking, values assigned to those variables nor past variables tested. Simple heuristics fixing the order in which those tasks are performed can lead to large efficiency improvements. If all solutions have to be found (i.e., exhaustive examination of the search space), then it makes sense to try first the paths that are more likely to fail, since this leads to an early pruning of the search space. If only one solution is needed, then trying first the paths that are more likely to succeed, leads faster to a solution. Thus, though the general heuristic is to assign consecutively variables related by explicit constraints, when searching for all solutions, variables with small domains and related by many explicit constraints should be assigned first, whereas when looking only for one solution, variables with large domains and related by few constraints should be assigned first. Similar opposite rules apply for value assignment: if all solutions are required, assign the most constrained (rare) value first, otherwise the less constrained. For testing past variables a good heuristic is to start with the most constrained past variable, whereas for the backtrack order itself, starting with past variables that have unsatisfied constraints with the current variable is usually best.

Besides these ordering techniques there are some theory-based heuristics based on a characterization of CSP using its dimension (number of variables), the cardinalities of the domains, and the relation matrix (Nudel 1983). Finally, some heuristics are based on the idea of solving a CSP modified such that it becomes backtrack-free and using the solution of the "relaxed" version to guide the search in the original problem space. These two last classes of heuristics do not always improve the overall efficiency, since their application itself is computationally expensive. Good heuristics are likely to be domain specific. Particularly in the

case of spatial constraints taking the rich structure of space into consideration
is the key to efficient algorithms as we will see in the next section.

5.4 Exploiting the structure of space

5.4.1 Combined structure of topological and orientation relations

In sections 4.3.3 and 4.4.2, we introduced the structure of the topological and
orientation relations, respectively (see figures 4.10, p. 38, and 4.14, p. 43). Since
we use pairs of topological and orientation relations to represent relative positions, it is interesting to look at their combined structure. As suggested by the
results of the combined composition of relation pairs in section 5.2.3, the combined structure is highly constrained. Since we are merging two "orthogonal"
dimensions, it is not possible to visualize the structure in a plane. Thus, we give
a simplified overview of level 3 orientations and linearly adjacent topological
relations in Figure 5.10, and a partial detailed view of two neighboring level 2
orientations and all 8 topological relations in Figure 5.11. As in Figure 4.10 arcs
between nodes denote neighboring topological/orientation pairs. In the simpli-

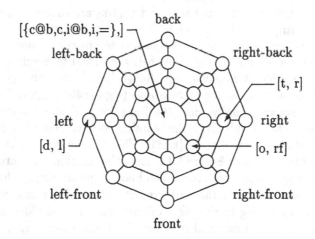

Figure 5.10: Combined structure of topological and orientation
relations (overview)

fied visualization the containment and equality topological relations have been
merged in a single node. This, of course, is not consistent with the fact that
i@b and c@b are oriented as opposed to c, i, =. Since i@b and c@b are no direct neighbors themselves, drawing an additional ring in the overview structure
would suggest non-existing neighborhoods between, e.g., [i@b,b] and [c@b,lb].
This is precisely the reason for the 3-D view shown in Figure 5.11. Figure 5.11
is intended to be seen as a perspective view of a side cut of the 3-D visualization

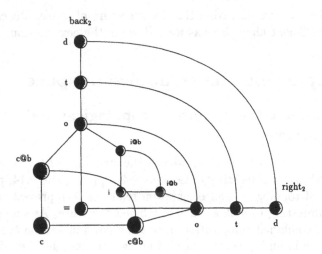

Figure 5.11: Combined structure of topological and orientation relations (detail)

of the combined structure. In order not to clutter the figure unnecessarily, only the two neighboring orientations back$_2$ and right$_2$ are shown, together with all the links connecting neighboring relation pairs. The c, =, and i nodes are in the middle of the structure, because, according to our conventions, they are not oriented. The oriented nodes (e.g., [i@b,b$_2$]) are linked to neighboring topological nodes of the same orientation (e.g., [o,b$_2$], [i,b$_2$]), and to equivalent nodes of same topology in neighboring orientations (e.g., [i@b,r$_2$], [i@b,l$_2$]).

The figures above actually omit the links for a class of neighboring relation pairs, those that result from the simultaneous change of topological and orientation relations as in [d,f] → [t,rf]. This can be seen in the alternative overview structure shown in Figure 5.12. Here the relation pairs are represented by areas, and neighboring areas mean neighboring pairs. Except for the center, which conveys information on the containment and equality relations, each of the eight areas corresponding to the eight orientations is subdivided in three slots corresponding to the symmetrical topological relations overlaps, tangent and disjoint. The figure also shows that the mentioned simultaneous changes are singular borderline cases, which are omitted from the explicit link representation to improve its readability.

5.4.2   Abstract maps

For some types of spatial reasoning it is advantageous to use internal data structures that inherently reflect as much as possible the properties of the represented domain. The structures introduced in sections 4.3.3 and 4.4.2, as well as the combined structure introduced in the previous section, reflect indirectly the structure of space through the structure of the relational domains used to de-

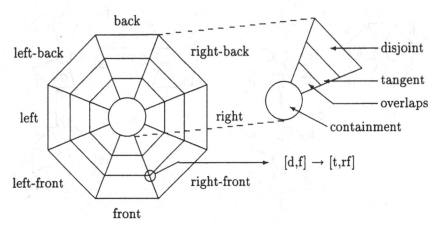

Figure 5.12: Alternative overview structure

scribe it. One way of exploiting these structures is introducing "abstract maps". Abstract maps contain for each object in a scene a data structure with the same neighborhood structure as the domain required for the task at hand.

For example, to model a change in point of view in a scene, which affects only orientation relations, we use a data structure called *ron* (for "relative orientation node"), with the same circular structure of the orientation domain. A *ron* has slots for each orientation, connected to form the same type of circular structure as the physical orientation domain. The orientation relation between two objects is modeled by creating a bidirectional pointer from the slot of the implicit orientation of the reference object to the corresponding slot of the "inverse relation" in the primary object.

A change in point of view, which affects relations with an explicit deictic type of reference frame, can be easily accomplished diagrammatically by "rotating" the labels of the orientation with respect to the intrinsic one. In the example shown, A is to the left of B (A [d,l] B) as seen from the original point of view (Figure 5.13a). If the point of view changes to be the one depicted in Figure 5.13b, A will rather be considered to be left-back of B as indicated by the rotated labels. Note that all relations of a given object to others in the scene are updated simultaneously while keeping topological relations consistent. So, while B is to the back of C in Figure 5.13a, their relationship is correctly relabeled in Figure 5.13b to be "B is right-back of C". Furthermore, this can be done "in parallel" for all objects in a scene.

The general data structure with the same neighborhood structure of the combined topological and orientation relations is called *rton* (for relative topological and orientation node).[6] It is important to remember, that rtons represent the neighborhood structure of the relations being used to represent space and *not* the metric of space itself. In particular, even though we tend to *depict*

---

[6]These structures were called *rpons* in previous work—see footnote in section 4.1.2 on page 28.

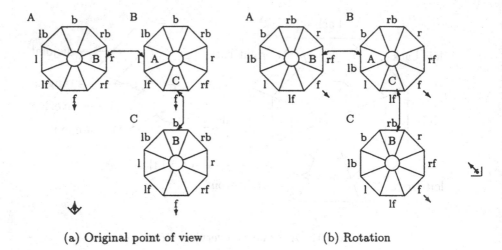

(a) Original point of view   (b) Rotation

Figure 5.13: Change in point of view as simultaneous rotation of labels

rtons in configurations that resemble the arrangement of the scene being represented (see, e.g., Figure 5.13), every topological deformation thereof would be an equally valid visualization. That is, "internally" only the connections between slots carry information.

A further example of the direct use of these structures can be found in the simultaneous composition of relations. While the table-lookup method discussed in section 5.2 is certainly more efficient for single compositions, it requires $n * m$ lookups and as many union operations for the composition of under-determined relations consisting of disjunctions of n and m primitive relations each, as the following code fragment from Allen (1983) shows:

```
Constraints (R1, R2)
  C ← ∅;
  For each r1 in R1
    For each r2 in R2
      C ← C ∪ T(r1, r2);
  Return C;
```

(Where T(r1, r2) returns the table entry for the composition of r1 and r2.) For example, if the composition of the two sets of relations {[d,l], [t,l]} and {[d,b], [d,lb], [d,l]} has to be computed, the six table lookups (and six set union operations) shown in Table 5.7 are needed. Given that all resulting compositions as well as most typical complex relations are connected (neighbors), using an rton to record partial results does all the required operations by virtue of its built-in structure (Figure 5.14).

It must be pointed out, that our representation of positions in 2-D is analogical only with respect to some properties of the relational domain (the neighborhood of relations) and not in general (Sander 1991). Consider what happens

$$
\begin{aligned}
[d,1]/[d,b] &= \{[d,1],[d,1b],[d,b],[t,1],[t,1b],[t,b]\} \\
[d,1]/[d,1b] &= \{[d,1],[d,1b]\} \\
[d,1]/[d,1] &= \{[d,1]\} \\
[t,1]/[d,b] &= \{[d,1b],[d,b],[t,1b],[t,b]\} \\
[t,1]/[d,1b] &= \{[d,1],[d,1b]\} \\
[t,1]/[d,1] &= \{[d,1]\}
\end{aligned}
$$

$$\cup \ \{[d,1],[d,1b],[d,b],[t,1],[t,1b],[t,b]\}$$

Table 5.7: Piecewise composition

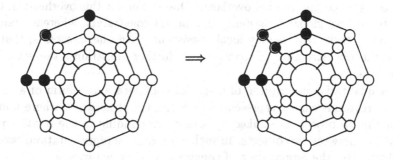

Figure 5.14: Using rtons to record partial results

if an object in a qualitatively represented scene is moved to a different position. This can lead in the worst case to a revision of all relations in the net referring directly or indirectly to that object. In a depictorial representation there is no need for such an update, because the new position of the object in the scene is reflected by the new position of the corresponding element in the representing medium. The point is: There is no need to update relations, because *no relations are represented at all*. Rather, whenever necessary, for example, when a query about the position of an object is done, an inspection process must be started to determine an appropriate way to describe it. Such an inspection process is by no means trivial. It requires establishing a point of view, choosing a reference object, a positional relation of an appropriate granularity level, and so on.

Thus, there is a tradeoff between maintaining a high level qualitative representation and maintaining a low level depictorial or diagrammatical representation. Paivio (1983) postulates in his dual-coding hypothesis the need for both forms of representation, since they complement each other. The visualization mechanisms discussed in section 6.2 can be used as part of an imaging process resulting in a depictorial representation on which some operations can be performed more efficiently than on its qualitative counterpart. This interim representation can be inspected to obtain a new set of relations describing it. However, because of the under-determined character of qualitative representations, this process may lead to serious information decay, unless other sources of knowledge or contextual information are available.

5.4.3   Propagation heuristics

Section 5.3 discussed general techniques for solving constraint satisfaction problems. While all spatial reasoning tasks in a qualitative representation can be formulated as constraint satisfaction problems, the general techniques incur an unnecessary computational overhead. One reason for this overhead is, that they try to achieve global consistency among all constraints, whereas many spatial reasoning tasks require only local consistency. Another reason is, that they ignore the rich structure of space, which further constraints the set of possible solutions.

A qualitative description of a spatial configuration in form of a set of positional relations can be represented as a constraint network, where nodes corresponding to objects are linked by arcs corresponding to the relative positional relations between two objects. In such a qualitative representation, two common tasks require the propagation of constraints in the network:

- Checking the consistency of the relations among adjacent objects, for example, after insertion of a new relation or deletion of a previously assumed relation.

- Establishing the relation between two objects not explicitly related by a positional relation. This requires finding a "path" between the two nodes, and computing the composition of the intermediate relations.

In the following subsections, we will describe algorithms for these two tasks that can be characterized by the following global design decisions:

- The structure of space should be taken advantage of at (at least) two different levels. On one level, the physical adjacency of objects limits the number of explicit positional relations among them.[7] On another level, the neighborhood structure of the topological and orientation domains reduces the number of alternative assignments.

- Determining the closure of the set of spatial relations is *not* a primary concern. In particular, no complete forward-checking algorithm should be used. Rather only a local consistency check should be done on insertion, and path-finding should be done "on-demand".

- The hierarchical and functional decomposition of space should be used to bound propagation.

- A weighting of positional relations according to their information content should be used to avoid "information decay" in the network due to the propagation of weak relations.

Data structures

In a constraint network implementation, data structures are needed for nodes (corresponding to objects), links (corresponding to sets of possible relations between objects), the global network (corresponding to a scene), and the composition tables (describing how to compose two relations). Furthermore, one or more auxiliary structures such as queues and stacks might be needed to manage the order of propagation. We discuss several alternative implementations for these structures, starting with the more common ones, and describing then our proposal as modifications thereof.

Sets of relations (relsets, for short), corresponding conceptually to disjunctions of possible relations between two objects, are ubiquitous in the algorithms to be described. Disjunctions are a way of expressing uncertainty about the "real" relation between the objects. Whenever the number of different relations is relatively small, as is always the case with qualitative representations, a bit-string representation of sets is the usual choice. A bit-string is a vector of binary elements, each associated with one of the elements of the set, such that if a bit in on (=1), the corresponding element is in the set denoted by the bit-string. The advantage of this representation is that all set operations can be mapped on binary operations on the bit-strings, which can be assumed to take constant time. For example, the union of two sets can be obtained by "oring" the corresponding bit-strings.

Relsets are the elements of the two-dimensional arrays used to represent the global network and the composition tables. The complexity of these structures

---

[7]As we will see in section 6.2 other, non-spatial factors such as functional clustering also limit the number of objects explicitly related.

varies, depending on the number of objects in the network, the number of relations to be composed, and the type of application.

In the simplest case a quadratic array of $n * n$ objects (initialized to the universal relset—containing all relations— for all $i \neq j$, and to = for all $i = j$), and
a quadratic array of $r * r$ relations (containing at $[r_i, r_j]$ the relset corresponding
to the r_k between A and C, if $A[r_i]B \wedge B[r_j]C$) will do. If many objects are available, a triangular matrix (implemented, for example, as a vector of vectors) can
be used to store the links between all objects o_i and o_j such that $i < j$, according
to some total order on the indices labelling the objects. This assumes that the
entry for $[o_j, o_i]$ can be computed by inverting the relations in $[o_i, o_j]$.[8] Likewise,
the strong symmetries in the composition of relations that have been discussed
in section 5.2 allow for more compact representations of the composition tables.

As we will see below, there are several reasons why the simple data structures
suggested up to now for nodes and links are not sufficient (nodes are represented
only implicitly as indices into the network table!). Besides assigning names to
objects, the parent object of the given object, as well as a list of other objects for
which this object is the parent object, should be kept explicitly. Furthermore, as
will be mentioned in the discussion, many applications might need to incorporate additional spatial (size, shape) and non-spatial (color, function) knowledge.
Thus, an object-oriented approach with classes, and instances thereof representing objects, seems to be called for. As for links, representing relsets alone does
not suffice, if we want to be able to retract relations derived through constraint
propagation when we delete the original relation. Rather, we have to provide
means of maintaining justifications (see "Deleting relations" below).

A more fundamental question is, if we want to more fully exploit the structure
of the topological and orientation domains using rtons and explicit links between
them to represent relsets instead of bit-strings. As explained in the previous section a relation between two objects is modeled by creating a bidirectional pointer
form the slot of the (implicit) relation of the reference object to the corresponding slot of the "inverse" relation of the primary object. This has the advantage of
allowing some operations to be performed "diagrammatically", that is, through
propagation of tokens in the rton structures, automatically enforcing the neighborhood structure of space. However, the overhead of maintaining full rtons for
each object with typically few links between them might prevent us from using
this approach. A compromise is to use a regular constraint network for "long
term storage", creating interim diagrammatical representations of parts of the
scene to perform those operations that require it. This is, on an abstract level,
similar to the envisionment process proposed in hybrid propositional/depictorial
systems (Pribbenow 1990; Chandrasekaran and Narayanan 1990).

"Binary" rtons can also be used as a direct replacement of bit-strings in
relsets. That is, the nodes in the rton corresponding to the positions between two
objects are set to on, while all others are off. While a software implementation of

---

[8]"Inverting" means here the positional relation obtained when exchanging primary and
reference objects under a constant frame of reference. This might not always yield useful
relations, particularly, when objects of different sizes—and accordingly of varying acceptance
areas—are involved.

such a data structure turns out to be less efficient than bit-string set operations, specialized hardware could provide constant time operations such as "rton-oring" for computing the union of two relsets, but also for more complex operations such as finding the composition of orientations by marking all positions on the "shortest path" between positions set by the union of the two original relsets.

Inserting new relations

Inserting a new relation between two objects[9] affects not only those two objects but might yield additional constraints on the relations between other objects in the scene through constraint propagation. Allen (1983) introduced an algorithm for updating an interval-based temporal network that is based on constraint propagation. This algorithm effectively computes the closure of the set of temporal assertions after each new insertion. Even though we said at the beginning of this section that we do not need to compute the full closure of our set of positional relations, it is useful to look at Allen's original algorithm and explain then the modifications we propose.

Figure 5.15 shows a corrected version of Allen's original algorithm broken up in three parts for better readability.[10] The first step just adds the new relation R_{ij} between i, j by intersecting the new relset with whatever relset was known before (this is here the universal relset, in case no relation had been previously inserted, which behaves as "identity" for the set intersection operation). In case this intersection results in a more constrained relset, the nodes i, j are placed in the queue for propagation.[11]

The computation of the closure is started by calling COMPUTECLOSURE, which calls PROPAGATE as long as it has entries in the queue. PROPAGATE does the main work by propagating the effects of the new constraint to "comparable" nodes (for now, assume all nodes in the network to be comparable). This is done by determining if the new relation between i and j can be used to constrain the relation between i and other nodes, or between those other nodes and j (Figure 5.16). If one of these relations can indeed be constrained, then it is placed in the queue for further propagation. Furthermore, contradictions, characterized by an empty resulting relset, are signaled if found in this process. Contradictions will normally trigger a constraint relaxation process as described in section 5.4.4.

To illustrate the propagation algorithm, we use the following set of initial relations, which have already been transformed to a canonical frame of reference:

1. `<T, [t,f], F>`

---

[9] Recall that we assume the relation has been transformed to the canonical implicit frame of reference.

[10] This resembles the version of Allen's algorithm given by Vilain et al. (1986). The original version in (Allen 1983) contains several typographical errors, that were corrected in later publications. The corrections were independently suggested by Bergler (1990), who also re-implemented Allen's algorithm in Scheme.

[11] Note that, because the new link is obtained by intersection with the old one, it suffices to test if the new one is different from the old one, which might be cheaper.

```
To  Add R_ij
   begin
      Old ← N(i,j);
      N(i,j) ← N(i,j) ∩ R_ij;
      If N(i,j) ≠ Old
         then put pair <i,j> on Queue;
      Nodes ← Nodes ∪ {i,j};
   end;

To  ComputeClosure
   While Queue is not empty do
      begin
         Get next <i,j> from Queue;
         Propagate(i,j);
      end;

To  Propagate i,j
   begin
      For each node k such that Comparable(i,k) do
         begin
            New ← N(i,k) ∩ Constraints(N(i,j),N(j,k));
            If New = ∅
               then Signal contradiction;
            If New ≠ N(i,k)
               then add <i,k> to Queue;
            N(i,k) ← New;
         end;
      For each node k such that Comparable(k,j) do
         begin
            New ← N(k,j) ∩ Constraints(N(k,i),N(i,j));
            If New = ∅
               then Signal contradiction;
            If New ≠ N(k,j)
               then add <k,j> to Queue;
            N(k,j) ← New;
         end;
   end;
```

Figure 5.15: Allen's propagation algorithm (adapted from Vilain et al. 1986, p. 375)

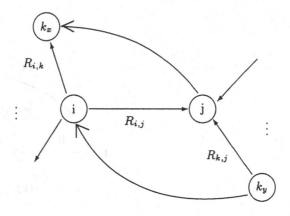

Figure 5.16: Propagation algorithm (visualization)

2. <S, {[d,f],[t,f]}, T>

3. <W, {[d,l],[d,r]}, T>

These relations might correspond to a spatial description such as the following:

1. *The table (T) is at the window (F).*

2. *In front of it there is a chair (S).*

3. *A bookcase (W) is next to it.*

Adding R_{TF} creates the first link in the network (Figure 5.17a). Adding R_{ST} triggers the computation of R_{SF} through composition of R_{ST}/R_{TF} (Figure 5.17b). Adding R_{WT} leads to the computation of R_{WF}, and, if we allow inverting links, to $R_{WS} = R_{WT}/R_{TS}$ (Figure 5.17c). Note, that the resulting relations in this last case are rather unspecific, because we do not know if the bookcase (W) is left or right of the table (T). Now suppose such information becomes indirectly available through a statement such as *"The bookcase (W) is to the right of the chair (S)."*, i.e., $R_{WS} = [d,r]$. The intersection with the previously computed relset [{d,t},{l,lb,b,rb,r}] results again in [d,r]. The propagation algorithm computes new values for $R_{WT} = R_{WS}/R_{ST}$ and $R_{WF} = R_{WS}/R_{SF}$ both equal to [{d,t},{r,rf,f}]. The intersection with previously computed relsets leads finally to $R_{WT} = [d,r]$ and $R_{WF} = [{d,t},{r,rf,f}]$ (Figure 5.17d).

Allen (1983) and later Vilain et al. (1986) showed the algorithm to run in polynomial time w.r.t. the number of intervals in the database. $O(n^3)$ set operations are required in the worst case for the algorithm to run to completion for n intervals, because there are at most n^2 relations between n intervals, each of which can only be non-trivially updated (and correspondingly entered on the queue for propagation) a constant number of times (each update removes at

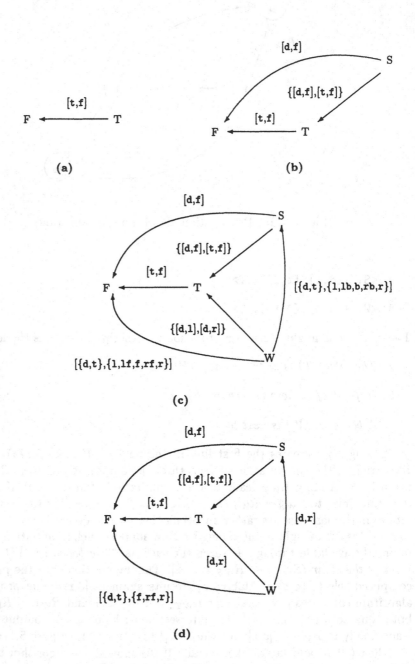

Figure 5.17: Propagation algorithm (example)

least one of 13 possible relations). In turn each of the $O(n^2)$ propagations requires $O(n)$ set operations resulting in the said $O(n^3)$ set operations (which in a bit-string implementation of sets can be assumed to take constant time each).

The predicate "comparable" is one place in the algorithm that allows us to introduce modifications. Allen himself uses it to control propagation by introducing "reference intervals" to cluster sets of fully connected intervals, and defining comparable to be true only if the intervals share a reference interval, or one is the reference interval of the other. Similarly, we can limit the propagation by assigning parent objects (i.e., those resulting from hierarchical decomposition through containment) or functional clusters (for example, "dining-table-group" consisting of table, chairs, etc.) as "reference objects". However, we do not require full connectivity of objects with a common reference object. In most cases, it suffices if all objects are related to at least the central object in the group (e.g., the table in the dining-table-group).

The next modification of the algorithm is not so straightforward. We want to take into account the fact that the information content of positional relations is not homogeneous. By information content we mean how much a relation constrains the relative position of objects. For example, a level 3 orientation is more constraining than a level 1 orientation, and t is more constraining than d. The specificity of a positional relation is, strictly speaking, a function of the relative size of the corresponding acceptance area. However, establishing those areas is more involved than what is actually needed to differentiate among the relevant specificity classes. Thus, we use the "degree of coarseness" (doc), a number derived from the number of options left open by a relation (and confirmed by the "coarsening" factor of the resulting compositions), as a converse approximate measure. Small doc-values correspond to more constraining relations, while large doc-values correspond to less constraining relations.

characteristics			rel	doc
topological			=	1
		size&shape restriction	c, i	4
	oriented		i@b, c@b	2
		boundary contact	t	3
			o	4
			d	8

Table 5.8: Degree of coarseness of topological relations (in the context of positional information)

Table 5.8 shows the degree of coarseness of topological relations in the context of positional information together with the factors contributing to the corresponding docs. = has the lowest doc, because it constrains the relative position

of two objects to an unique value; c and i are not oriented but constrain the
position of one of the objects to be within the boundaries of the other; t and o
demand boundary contact between the objects, thus restricting their positions,
whereas d has the highest doc and is useful only when used together with an
orientation;[12] finally i@b and c@b are quite specific because they are oriented,
demand boundary contact and have the size and shape restriction of the containment relations.

characteristics			rel	doc
orientation	level 3	corners	lf_3, rf_3, lb_3, rb_3	1
		sides	l_3, r_3, f_3, b_3	3
	level 2		l_2, r_2, f_2, b_2	4
	level 1		l_1, r_1	8
			f_1, b_1	8

Table 5.9: Degree of coarseness of the most common orientation
relations (in the context of positional information)

Table 5.9 lists the degree of coarseness of the most common orientation relations. It corresponds roughly to the number of sectors of the common level
(Figure 4.15 on page 44) covered by each of the relations. For the case of extended objects, a further distinction between more specific corner orientations
and less specific side orientations can be made, as shown here for the third level
(see section 4.4.4, particularly Figure 4.21). The doc of a topological/orientation
pair is the sum of the docs of its components. The doc of a relset is the sum of
the docs of the set members.

The doc-values are used to control propagation in the following way: While
adding a new relation (Figure 5.18), its doc is compared against a pre-defined
constant maxdoc, above which a relation is considered too unspecific to be worth
adding. The value of maxdoc is application dependent. Setting it to half the sum
of the docs of all possible relations at the granularity levels allowed has proven
to be a useful heuristic. If the doc of the relset is below or equal to maxdoc, then
it is inserted by combining it with the previously known relset. COMBINE can be
assumed for now to be equivalent to set intersection (it will be modified below).
If the new combined relation is different from the previously known, it is placed
on the agenda for further propagation. An agenda is a data structure to keep
track of what to do next based on some sort order. It is usually implemented
as an ordered list of queues. In this case, we use the doc of the relation to be
propagated as entry key, allowing us to propagate first more specific relations,
i.e., those with lower docs. Relations with equal docs are processed in a first-in-
first-out manner.

---

[12]The unspecificity of d should not be overrated: In most applications, the distance between
objects is bounded by the extension of the parent object.

```
To  ADD R_ij
  begin
    if doc(R_ij) > maxdoc
       then exit;
    Old ← N(i,j);
    N(i,j) ← Combine(N(i,j),R_ij);
    If N(i,j) = ∅
       then Signal contradiction;
    if N(i,j) ≠ Old
       then add <i,j,doc(N(i,j))> to Agenda;
    Nodes ← Nodes ∪ {i,j};
  end;
```

Figure 5.18: Adding a new relation

```
To  COMPUTEEFFECTS
  While Agenda is not empty do
    begin
      Get next <i,j,d> from Agenda;
      If d ≤ maxdoc
         then Propagate(i,j);
    end;
```

Figure 5.19: Computing the effects of a new added relation

```
To  PROPAGATE i,j
  begin
    For each node k such that Comparable(i,k) do
      begin
        New ← Combine(N(i,k),Constraints(N(i,j),N(j,k)));
        If New = ∅
          then Signal contradiction;
        If New ≠ N(i,k)
          then add <i,k,doc(New)> to Agenda;
        N(i,k) ← New;
      end;
    For each node k such that Comparable(k,j) do
      begin
        New ← Combine(N(k,j),Constraints(N(k,i),N(i,j)));
        If New = ∅
          then Signal contradiction;
        If New ≠ N(k,j)
          then add <k,j,doc(New)> to Agenda;
        N(k,j) ← New;
      end;
  end;
```

Figure 5.20: Weighted Propagation

The effects of new relations added to the network can be computed by calling COMPUTEEFFECTS (Figure 5.19), which fetches the first entry from the agenda, double checks its doc to be below the limit (this is necessary because PROPAGATE also adds to the agenda, and could also be used for dynamic control through changing limits), and calls PROPAGATE.

PROPAGATE (Figure 5.20) is essentially the same as in the original algorithm (Figure 5.15), except for COMBINE and the use of an agenda instead of a queue. Note, that the propagation algorithm assumes all relations, including semantically similar relations at different levels of granularity (e.g., b_2 and b_3), to be mutually exclusive. Knowing about the hierarchical structure of the relational domains used, allows us to extend the algorithm to succeed in cases where the original algorithm would fail, and signal a contradiction. For example, the intersection of a set containing only $[t,l_2]$ and a set containing only $[t,l_3]$ shouldn't be "empty", signaling a contradiction, but rather lead to preferring $[t,l_3]$. This can be done by looking at the range representation of l_2 and l_3, instead of viewing them as unrelated relations. Furthermore, the intersection of sets containing neighboring relations, such as for example $\{[t,l_3]\}$ and $\{[t,lb_3]\}$, should not be "empty", but rather lead to a coarser relation such as $[t,l_2]$. These extensions are implemented by the modified COMBINE in Figure 5.21, which checks to see if subsumed or neighboring relations are available, if the regular intersection of two relsets is empty. In the first case, the subsumed (more specific) relation is placed in the new combined set. In the second case, the "next common coarse relation" (nccr) is added to the combined set. This is a form of constraint relaxation (see also section 5.4.4) embedded in the propagation process.

```
To  COMBINE R1, R2
    begin
      temp ← R1 ∩ R2;
      if temp ≠ ∅
        then Return temp;
      C ← ∅;
      For each r1 ∈ R1
        For each r2 ∈ R2
          begin
            if subsumes(r1,r2) then C ← C ∪ r2;
            if neighbor(r1,r2) then C ← C ∪ nccr(r1,r2);
          end;
      Return C;
    end;
```

Figure 5.21: Combining two relsets

Continuing our example, assume the initial relation between W and T to be $\{[d,l_2],[d,r_2]\}$, and suppose that we learn later that $R_{WT} = [d,rf_3]$. Instead of returning an empty set as an intersection of the two sets would, COMBINE recognizes that r_2 subsumes rf_3 and returns $[d,rf_3]$ as result. After propagation, the relations between W and F, and between W and S are constrained to be $\{[d,f],[d,rf]\}$, and $[\{d,t\}, \{b,rb,r\}]$, respectively.

As a consequence of the modified COMBINE, New ≠ N(i,k) in PROPAGATE being true does not imply New ⊂ N(i,k). However, the worst case complexity analysis of the original algorithm is not affected by this change, because every non-trivial update of a relset either removes at least one relation (intersection, original algorithm), or replaces one relation by a subsumed one, or one or more fine relations by a coarser one. The average case performance is greatly improved by the modifications described, particularly by the hierarchical decomposition, and by the preferential propagation of specific relations.

Deleting relations

Deleting relations between two nodes is not just a matter of removing a link from the data structure representing the network. The consequences of the propagation of the constraint now being deleted must be taken back as well.

This requires a further modification of the insertion algorithm described in the previous section to maintain justifications for derived constraints. Instead of just modifying a link to contain the new constrained relset, we also record the link whose propagation led to the new constraint. In general, a justification is a list of the links that were used to derive the relation of the new link. To allow for multiple derivation paths, usually a list of justifications is maintained. At the same time a link has pointers to those links that it in turn served in deriving in a so called justificands list. Relations originally entered by the user (considered as "premises") have an empty justification list. This information is usually maintained in a separate "dependency network", where the links of the constraint network are the "nodes",[13] and the arcs connecting them represent the dependency structure. Figure 5.22 shows a typical graphical representation of dependency networks, and illustrates the terminology introduced above (empty justifications marking premises are shown as solid rectangles). The dependency network can also serve as direct indexing and retrieval mechanism for sets of consistent relsets.

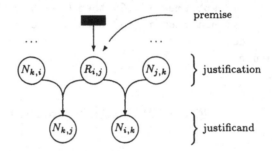

Figure 5.22: Graphical representation of dependency networks

The process operating on a dependency network is called "reason maintenance". Because they can operate independently from the problem solver (the constraint propagation algorithm, in our case), reason maintenance algorithms and the corresponding dependency networks have been developed as separate systems called "Reason Maintenance Systems" (RMS). RMSs were introduced in the late 70s in the context of computer-aided circuit analysis by Stallman and Sussman (1977), and first studied as independent systems by Doyle (1979).[14] Further variants were later developed by McAllester (1980), de Kleer (1986), and others.

Based on the dependency information recorded during the propagation of constraints, the RMS establishes which nodes of the dependency network are affected by the deletion of a link in the constraint network. Figure 5.23 shows the dependency network for the example from the previous section, after the

---

[13]Note, however, that each particular relset R_{ij} between objects i and j is recorded as a separate node of the dependency network.

[14]They were originally called "Truth Maintenance Systems", a somewhat misleading term, still often found in the literature. We prefer the name "Reason Maintenance Systems" following McDermott (1983).

first set of relations (the "premises" R_{FT}, R_{ST}, R_{WT}) and its consequences (R_{SF}, R_{WF}) have been established.[15] The one as superscript indicates that this particular relset is the first one assigned to that link. Figure 5.24 shows the

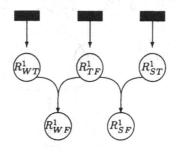

Figure 5.23: Example dependency network (initial relations)

next step in which R_{WS} is added as a premise, resulting in the derivation of new values for R_{WT}^2 and R_{WF}^2 (indicated by the two as superscript). Note that the old values R_{WT}^1 and R_{WF}^1 are cancelled out. Now suppose we originally got the

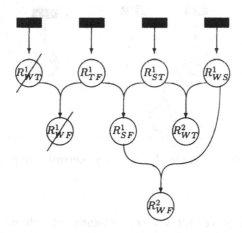

Figure 5.24: Example dependency network (propagation)

wrong description and are forced to delete the relation between S and T. In that case, all the consequences derived by using that relation must be removed as well, as shown in Figure 5.25. Unfortunately, we are left only with the relations R_{TF}^1 and R_{WS}^1, even though we had previously the relations R_{WT}^1 and R_{WF}^1 that did not depend on the erased relation. The RMS mechanism, however, does usually reconsider the evidence in support of all nodes based ultimately on premises, when other values are erased, and would restore in this case the

---

[15]For better readability, we are omitting here the propagation through inverted links (e.g., $R_{WS} = R_{WT}/R_{TS}$).

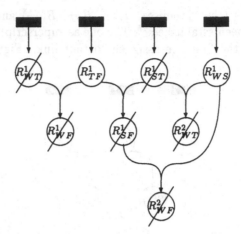

Figure 5.25: Example dependency network (deletion)

old values for R_{WT}^1 and R_{WF}^1 (Figure 5.26). Thus, maintaining dependencies

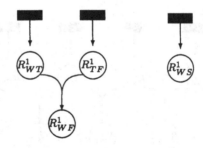

Figure 5.26: Example dependency network (old context restored)

not only allows us to retract the consequences of deleted relations, but also helps us to avoid repeated computations. Of course, there is always a tradeoff between the overhead of caching inferences and the costs of recomputing them, particularly when restrictive propagation policies apply.

In what follows we give an outline of the algorithms making up the RMS mechanism. The information associated with a node of the dependency network is stored in a record structure called molec (Figure 5.27).

Figure 5.28 shows the extensions to the insertion mechanism required to record dependency information. The new link of the constraint network is given together with the links that led to its derivation as argument to a procedure that generates a new dependency network node and installs its justifications. The procedures ADD and PROPAGATE (figures 5.18 and 5.20, respectively) must be extended to include a line like, for example, ADDTONET(New, (N(i,j), N(j,k))).

```
molec: record
        name: string;
        relset: set of relations;
        supporters: list of molecs;
        justifications: list of lists of molecs;
        justificands: list of lists of molecs;
end;
```

Figure 5.27: Molec record structure

```
To ADDTONET New justification
  begin
    m ← molec(New);
    dd-install(m, justification);
    molec.supporters(m) ← some satisfied molec.justifications(m);
    molec.status(m) ← IN;
  end;

To DD-INSTALL m j
  begin
    For each s in j
        molec.justificands(s) ← molec.justificands(s) ∪ m;
    molec.justifications(m) ← molec.justifications(m) ∪ j;
  end;
```

Figure 5.28: Recording dependency information

Figure 5.29 shows the code that triggers reason maintenance upon deletion of a link in the constraint network, and the corresponding erasure of a node in the dependency network. The RMS procedure used here is similar to the "garbage collection" mechanism in LISP systems (Figure 5.30): The status and supporters of all molecs in the network are reset first, and then recomputed by a recursive mark procedure that—starting from the premises—finds all nodes with "well-founded" support. Nodes without such support are physically removed from the network by the CLEANUP procedure, which is assumed here to be a "system primitive", and not further detailed. While the procedure illustrates the basic idea, it is usually not very efficient, because all nodes in the network are visited (only once, but anyway).

Thus, other variants start from the erased node, and recursively mark all its justificands for erasure. However, because of the potential of circular depen-

```
To  ERASE m
   begin
     dd-erase(m,molec.supporters(m));
     rms;
   end;

To  DD-ERASE m j
   begin
     For each s in j
         molec.justificands(s) ← molec.justificands(s) \ m;
     molec.justification(m) ← molec.justifications(m) \ j;
   end;
```

Figure 5.29: Deleting dependencies

dencies, a somehow more involved procedure is needed (Figure 5.31). It resets the non-circular supporter/supportee links and assigns them a tentative status, assuming that all nodes marked inactive ($\neq$ {IN, OUT}) will ultimately be erased (OUT). If that happens not to be the case, i.e., if one such molec becomes active (IN), because of a valid justification, all its supportees must be reconsidered, because they might have been assigned a new status based on the assumption that the supporter be OUT. Details of this algorithm can be found in (Hernández 1984), where several variants were implemented based on ideas in (Charniak et al. 1980) and (Doyle 1979).

Path finding

Since we do not maintain full connectivity of all nodes in the network, an algorithm to find a path between two nodes in the network might be required, for example, to answer a question about the possible relative position of the corresponding objects. Once such a path has been found, computing the composition of relations along it yields the desired relation.

Again, we use the algorithm proposed by Allen (1983) as starting point and modify it according to the design decision stated at the beginning of this section. Allen's algorithm works by searching up the reference hierarchy until a path (or all paths) between the nodes are found. The search is done by straightforward graph search, except that at each step the path must be between a node and one of its reference nodes or be a direct connection. That is, the general form of the path is

$$n_1, n_2, \ldots n_k, n_{k+1}, \ldots n_m$$

with:

- for $1 \leq i \leq k - 1$, n_{i+1} is a reference node for n_i;

```
To  RMS
  begin
    For each m in Known-relations
        molecs-clobber(m);
    For each m in Known-relations
        If premise(m) then mark(m,0);
    For each m in Known-relations
        If molec.status(m) = false then cleanup(m);
  end;

To  MOLECS-CLOBBER m
  begin
    If molec.status(m) ≠ false then
      begin
        molec.status(m) ← false;
        molec.supporters(m) ← ∅;
        For each s in molec.supportees(m)
          molecs-clobber(s);
      end;
  end;

To  MARK m supps
  begin
    molec.status(m) ← IN;
    molec.supporters(m) ← supps;
    For each j in molec.justificands(m)
      If molec.status(m) = false ∧
        some molec.justifications(m) satisfied
        then mark(j,satisfied justification);
  end;
```

Figure 5.30: Reason maintenance (garbage collection version)

```
To  ERASE m
  begin
    *to-be-erased* ←  ∅;
    dd-erase(m,molec.supporters(m));
    molecs-clobber-status(m,0);
    molecs-assign-status(m,0);
    For each m in *to-be-erased*
        If molec.status(m) = OUT then cleanup(m);
  end;
```

```
To  MOLECS-CLOBBER-STATUS m level
  If molec.status(m) ∈ IN OUT then
    begin
      molec.status(m) ← level;
      molec.supporters(m) ←  ∅;
      For each s in molec.supportees(m)
        molecs-clobber-status(s,level);
    end;
```

```
To  MOLECS-ASSIGN-STATUS m level
  If molec.status(m) = level then
    If assign-tentative-status(m) = IN then
        begin
          For each s in molec.supportees(m)
              molecs-clobber-status(s, level + 1);
          For each s in molec.supportees(m)
              molecs-assign-status(s, level + 1);
        end;
        else *to-be-erased* ← *to-be-erased* ∪ m
```

```
To  ASSIGN-TENTATIVE-STATUS m
  begin
    supps ← some molec.justifications(m) satisfied;
    If supps ≠ ∅ then
        begin
          molec.supporters(m) ← supps;
          molec.status(m) ← IN;
        end;
        else molec.status(m) ← OUT;
  end;
```

Figure 5.31: Reason maintenance (multi-level version)

- n_k and n_{k+1} are connected explicitly;

- for $k + 1 \leq i \leq m - 1$, n_i is a reference node for n_{i+1}.

The same restriction imposed on the propagation of weak information while inserting new relations apply here while composing relations along the path found. The docs of relations provide additional criteria for selecting among many paths besides their length. This can be useful to avoid having to compute first the composition along all paths, and then intersect the results to obtain the most constrained one.

5.4.4 Constraint relaxation

Another closely related form of reasoning is given whenever we try to design a physical layout to fulfill a set of spatial requirements. Since these requirements are typically dictated by associated non-spatial (mostly functional) criteria, they might not correspond to any physical situation at all. In other words, the spatial constraints expressed by the relations might not be consistent with each other. This situation typically arises in the first stages of design, when functionally motivated spatial constraints are verbally formulated. Consider, for example, the following situation from urban planing: The area of an old airport is to be converted into a residential area and a business area. These areas should be as far apart from each other as possible. On the other hand, they both should be connected to the public transportation system by a single station. Table 5.10

Design constraint	Rel. notation	Icon
Residential and business areas should be disjunct	R [d] B	
Residential area should contain a public transportation station	R [{c, c@b}] P	
Business area should contain a public transportation station	B [{c, c@b}] P	

Table 5.10: Excerpts from design constraints

formulates these design requirements in terms of our topological relations, while Figure 5.32 summarizes them as constraint graph (again, icons derived from the structure in Figure 4.10 are used to represent sets of relations compactly). These constraints, taken together, are unsatisfiable, i.e., a propagation would yield empty solution sets. Given a set of unsatisfiable constraints the task of removing some of the constraints from the given set such that the resulting subset of constraints can be satisfied is called constraint relaxation. Of course, the resulting set of constraints should be as close as possible to the original set of

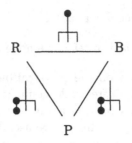

Figure 5.32: Constraint relaxation (example graph)

constraints. This problem is known to be NP-complete in the general case, because one would have to try to remove all possible combinations of subsets, check
if the resulting sets are satisfiable, and then somehow measure the distance of the
result to the originally given set in order to choose the best candidate among the
solutions found.[16] Thus, the usual constraint relaxation procedure corresponds
to an uninformed complete withdrawal of one of the constraints. Depending on
which constraint is retracted, this would mean giving up the desired separation
of the areas (noise, pollution in residential area) or the requirement of public
transportation. Table 5.11 summarizes the resulting unsatisfactory situations.

The problems with these "solutions" arise only because the internal relations
among constraints are being ignored. When confronted with unsatisfiable constraint combinations, humans do not generally reject categorically some of them,
but rather tend to find a suitable combination of weakened reformulations of the
original constraints. If the internal structure of topological relations is taken into
consideration, then the constraints can be weakened by including other neighboring relations in their disjunctive definitions, instead of retracting them as a
whole. This approach leads to the modified set of constraints in Table 5.12.

After propagation we obtain the intuitive solution in Table 5.13. That is, the
obvious way to share the public transportation station is allowing the areas to
overlap with one exit on each of the residential and business sides.

It is important to realize that this solution is obtained without further knowledge sources such as priority lists, because all constraints are weakened simultaneously. In the worst case, of course, this procedure must be repeated until
a solution is found. Furthermore, a priority list can still be used to improve
the search, for example, there is no real need to weaken R [{d}] B to R [{d,
t}] B. The combination of knowledge from more than one domain can be used
to improve the constraint relaxation process further. In the domain of topological relations, for example, if we know that the region of the business area
B is greater than the region of the station P, it makes no sense to weaken the
topological relation between B and P to allow B to be included in P. That is,

---

[16]For technical reasons, the standard algorithm for this kind of problem (see for example
Hertzberg et al. 1988) actually works the other way around (see Hernández and Zimmermann
1993, for further details).

Constraint withdrawn	Result after propagation	Graph
R [d] B	R [{o, c@b, i@b, =}] B	
R [{c, c@b}] P	R [d] P	
B [{c, c@b}] P	B [d] P	

Table 5.11: Uninformed withdrawal

Original constraint	Weakened constraint	Graph
R [d] B	R [{d, t}] B	
R [{c, c@b}] P	R [{c, c@b, o}] P	
B [{c, c@b}] P	B [{c, c@b, o}] P	

Table 5.12: Structured constraint relaxation

Solution	Interpretation	Graph
R [{d, t}] B	Residential and business areas may be adjacent (touch)	
R [o] P	Residential area "shares" a station (partially overlaps)	
B [o] P	Business area "shares" a station (partially overlaps)	

Table 5.13: Plausible solution

additional knowledge from another domain, in this case size, allows us to ignore the relations i and i⊘b without missing possible solutions to the given problem. Some topological relations such as d have only one neighbor (t), whereas others such as o have four (t, c⊘b, i⊘b, =). The question arises as to which neighbors should be included in the weakened constraint, if the domain structure is such that there are several neighbors. While additional knowledge such as the size of objects, as we just saw, can be used to decide which neighbors to include, the computational overhead of just including *all* of them is relatively low, because objects are usually related to only a couple of other adjacent objects, thus limiting the scope of propagation.

5.4.5 Coarse reasoning and hierarchical organization

The hierarchical structure of the topological and orientation domains was introduced in sections 4.3.3 and 4.4.2, respectively. Spatial reasoning can be done at coarser or finer levels of that structure, depending on the kind of information available. In particular, if only coarse information is available, the reasoning process is less involved than if more details are known. This is a very attractive aspect of our approach, which distinguishes it from other frameworks (e.g., using value ranges or confidence intervals) in which less information means more computation.

Abstract maps, for example, facilitate "coarse" reasoning by using relations that make less distinctions either to cope with missing information or to simplify reasoning. We could for example only know that the primary object is on the left and has contact to the reference object. In that case it suffices to distinguish between left and right in the orientation dimension, and between contact and no-contact in the topological dimension. In our analogical representation this corresponds to using levels of rtons with less subdivisions (in the visualization of rtons, discs with less subdivisions are superimposed on the original octagon).

5.5 Summary

We present a variety of mechanisms to reason with qualitative representations in general, and qualitative representations of 2-D positional information in particular. One of the simplest is transforming between reference frames. In addition to the intrinsic orientation of the parent object, the intrinsic orientation of the reference object (in the intrinsic case), the external factor (in the extrinsic case), or the point of view (in the deictic case) are needed to compute the distance that the labels of the orientation structure need to be rotated to obtain the implicit orientation.

The composition of spatial relations turns out to be strongly under-determinate, i.e., the resulting relation sets tend to contain too many alternative relations, if the topological and orientation domains are considered separately. Fortunately, the composition of positional information, i.e., of topological/orientation pairs, yields more specific results. Interestingly, only a fraction of all possible variations of relations actually occur as compositions, and all of them are connected. Also "fixed-points" are reached after a few iterations. Simple rules for the computation of composition can be given (for example, "shortest path" rule for orientations).

We give an overview of the solution techniques for the general constraint satisfaction problem available in the literature. Even though the solution techniques described represent large efficiency improvements over the obvious backtracking algorithm, they are "limited by their generality". That is, being general, domain-independent techniques, they ignore the structure of the relational domain. Thus, we show that taking the structure of the richly constrained spatial domain into consideration leads to more efficient algorithms.

One way of exploiting this structure is introducing "abstract maps". Abstract maps contain for each object in a scene a data structure with the same neighborhood structure as the domain required for the task at hand. A change in point of view, for example, can then be easily accomplished diagrammatically by "rotating" the labels of the orientation with respect to the intrinsic one. Another way is introducing heuristics to control the propagation of constraints by using the hierarchical and functional decomposition of space to limit constraints to physically adjacent objects. Also, a weighting of positional relations according to their information content is used to avoid "information decay" in the network due to the propagation of weak relations. Finally, we discuss a method for constraint relaxation that uses the structure of the relational domain to weaken constraints by including other neighboring relations in their disjunctive definitions, instead of retracting them as a whole. This approach leads faster to solutions of meaningfully modified sets of otherwise unsatisfiable constraints.

Chapter 6

Applications

What we have to learn to do we learn by doing.

Aristotle, *Ethica Nicomachea II, c. 325 B.C.*

Qualitative representations of space can be used in many application areas from everyday life in which spatial knowledge plays a role, particularly in those that are characterized by uncertain and incomplete knowledge. This is the case, for example, in the first stages of design, in which verbally expressed spatial requirements and raw sketches are very common. Examples of such domains are computer aided systems for architectural design and urban planning, geographical information systems, natural language information systems to give directions and for robot control. Qualitative representations have also been applied to the description of document layouts (Fujihara and Mukerjee 1991) (see section 8.1) and furniture layouts (Pfefferkorn 1975), a domain which we have been using as example in this book.

Rather than going into the domain-dependent details of particular applications, we will look into two essential tasks, which a representation of spatial knowledge must be able to deal with:

- Given a spatial configuration (either physically or through a verbal description) create a representation thereof (knowledge acquisition). A closely related task is to incorporate new knowledge into an existing representation (knowledge assimilation).

- Given an internal representation envision the corresponding spatial configuration. This "envisionment"[1] might be partial, as in the case of answering simple queries about the spatial configuration, or actually require the recreation of the scene being represented.

---

[1] In the qualitative physics literature, "envisionment" is used in the restricted technical sense of a network of all possible states and transitions for a system. Each path through that network is called a "history".

Both tasks rely on the reasoning mechanisms introduced in chapter 5 for maintaining the consistency of the knowledge base, deriving new or implied knowledge, etc.

Applications as different as natural language query systems (where both knowledge assimilation and knowledge use are done through natural language) and visual oriented systems (where input and output have a graphical format) have the same basic structure as can be seen in figures 6.1 and 6.2 respectively. The issues involved in the acquisition and assimilation of verbally expressed

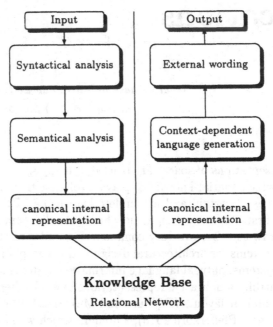

Figure 6.1: NL Query-System

knowledge and in the construction of natural language query systems have been extensively explored in the literature. Thus, in this chapter we emphasize the visual aspect as seems appropriate for spatial knowledge but not obvious in the case of qualitative representations. Section 6.1 describes the steps required to build a "cognitive map" from single views of a scene in a simplified setting. While what is being represented here are not positions but angles and wall sizes, the process illustrates how such a representation can significantly reduce the complexity of the procedure, while increasing its "cognitive plausibility". The visualization of office layouts discussed in section 6.2 is an example of the use of qualitative knowledge in the envisionment process.

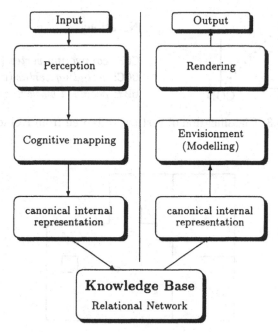

Figure 6.2: Visual oriented system

6.1   Building cognitive maps

As an example of how going from a quantitative to a qualitative framework might
be advantageous, we have been looking at the "Computational Theory of Cognitive Maps" of Yeap (1988).[2] Yeap's work emphasizes the acquisition of a "raw
cognitive map" from perceptual information (reduced for simplicity to a "$1\frac{1}{2}$-D
Sketch" containing the projection to two dimensions of all relevant surfaces in
a room). The goal is to obtain what Yeap calls a stable *absolute space representation* (ASR), i.e., an observer-independent description of the room, from
a series of observer-dependent views. Yeap's original algorithms rely heavily
on the classification of the observed vertices delimiting surfaces in *connecting*,
occluding and *occluded-by* (see Figure 6.3) and the heuristic that neighboring
vertices of the same type should be connected to form one of the surfaces con-
stituting the ASR. In a multi-step procedure, a list of vertices obtained from
the pre-processing module is classified in those forming groups corresponding to
surfaces and those that belong to occluded surfaces. Local and global heuristics
eliminate then spurious surfaces. The right hand of figure 6.7 shows the steps
of Yeap's algorithm, embedded in the general task of building cognitive maps
described below. A re-implementation of the system done by Hafner and Kobler

---

[2]Notice, however, that the spatial elements to be represented here are not positions but
angles and wall sizes. Thus, this section is not a direct application of the positional model
derived in previous chapters, only an application of the qualitative approach.

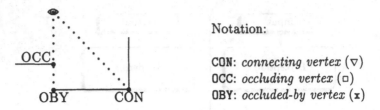

Notation:

CON: *connecting vertex* (∇)
OCC: *occluding vertex* ($\square$)
OBY: *occluded-by vertex* (x)

Figure 6.3: Classification of vertices (adapted from Kobler 1991)

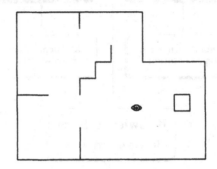

Figure 6.4: Layout of a room used as example (from Kobler 1991)

(1991) revealed serious deficiencies in the original algorithms. In particular, small variations of the observer's position lead in many cases to wrong results even for standard types of rooms. Kobler (1991) analyzes the causes for this behavior and describes improved algorithms that yield more stable representations. Although taking many cognitively plausible elements into consideration such as a short term restricted representation of the immediate surroundings, the algorithms work with exact quantities for sizes and angles of perceived surfaces. We propose using qualitative information instead. By qualitative information we mean *relative* sizes of surfaces to each other and to prototypical instances of selected domain classes, and *relative* orientation of surfaces to each other. Thus, what we do is "replace" the quantitative elements in Yeap's representation by qualitative ones. We then try to reproduce the functionality of the original model to find out where the strengths and weaknesses of this "representational shift" lie. This kind of representation, given suitable context mechanisms, actually simplifies the "recognition of places" problem (i.e., the matching of a perceived situation against those situations already known by the agent).

These extensions of Yeap's algorithms form the basis for an experimental system that we envision as follows: The system gets a 2-D scanned layout plan of a building (reduced to the "relevant" surfaces) and an initial position as input. Based on the layout plan, the first stage of processing generates the partial views—as seen by a hypothetical agent—thus simulating a robot's "perception"

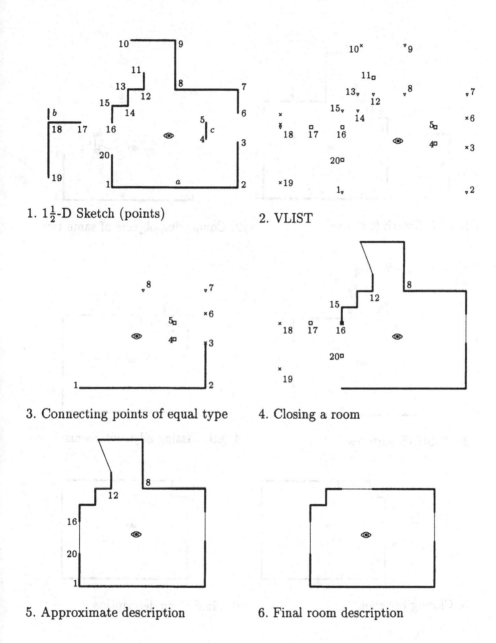

1. $1\frac{1}{2}$-D Sketch (points)   2. VLIST

3. Connecting points of equal type   4. Closing a room

5. Approximate description   6. Final room description

Figure 6.5: Yeap's algorithm for building cognitive maps (from Hafner and Kobler 1991)

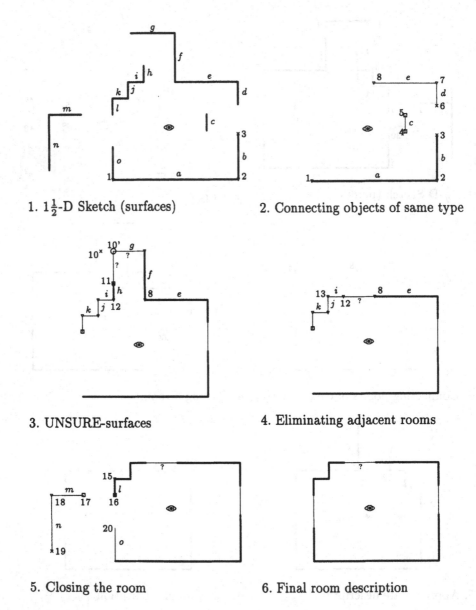

1. $1\frac{1}{2}$-D Sketch (surfaces)   2. Connecting objects of same type

3. UNSURE-surfaces   4. Eliminating adjacent rooms

5. Closing the room   6. Final room description

Figure 6.6: Modified algorithm for building cognitive maps (from Hafner and Kobler 1991)

while wandering through the building. Those partial views are used next to create a relative, egocentric representation of the perceived room, a process that includes the recognition of spatial boundaries and exit points and the position of other surfaces relative to those boundaries or exits. This *egocentric* representation, in which all information is relative to an agent, is just the first stage in the process of building a "cognitive map" (Hart and Moore 1973). In order to perform useful spatial reasoning, the agent must be able to transform this representation into non-egocentric forms: an *allocentric* one, in which spatial information is expressed relative to distinguished reference structures, and a *geocentric* one containing abstract topologic and metric relations in a coordinated system of reference frames. A common characteristic of all stages is

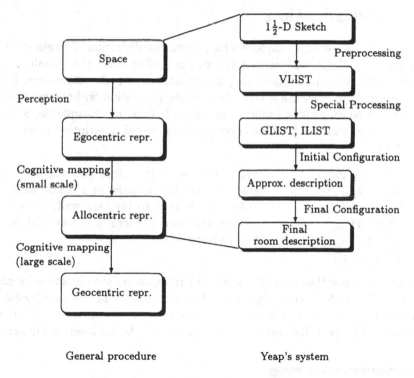

Figure 6.7: Building cognitive maps

the hierarchical clustering of relative information. In our setting, the most important kind of information going beyond single rooms is connectivity, which emphasizes that the way (including temporal order) in which we experience our environment strongly influences the representation we build thereof.

Parts of this system have been implemented. Hahn (1990) describes a reimplementation of Funt's retina (Funt 1980) in PC-Scheme, which can be used as a simple recognition mechanism, since the images consist mostly of straight lines and right angles. The program uses a scanned layout plan of a group of offices as an operating example and includes mechanism to recognize relevant features such as walls, doors and obstacles. Of course, classic pattern recognition algorithms are probably more efficient, but the diagrammatic method has the potential of being parallelized easily on very simple processors. Note that the layout plan is used as a cheap replacement for an actual perceptual system and is not available as such for internal processing (otherwise we would already have a spatial representation!). As was mentioned above, a re-implementation of Yeap's "Computational model of cognitive maps" has been done by Hafner and Kobler (1991) (see also Kobler 1991). It includes useful visualization of the process of building cognitive maps from lists of surfaces (which could be provided by Funt's retina).

6.2   Visualization

In this section we want to explore what, given the definition of qualitative representations in chapter 2, might seem an impossible task: the visualization of a given qualitative description, i.e., a reconstructive task. Obviously, human beings are able to visualize a verbally described situation without much effort. So, why is it so difficult to imitate this ability? Take for example the following set of relations (with a possible corresponding verbal description given to the right of it):

$$\langle \mathcal{T}[\{d, t\}, r] \mathcal{W} \rangle$$
$$\langle \mathcal{S}[t, f] \mathcal{W}_{\text{rear}} \rangle$$
$$\langle \mathcal{D}[\{d, t\}, r] \mathcal{S} \rangle$$
$$\langle \mathcal{D}[t, f] \mathcal{W}_{\text{rear}} \rangle$$
$$\langle \mathcal{D}[t, l] \mathcal{W}_{\text{right}} \rangle$$
$$\langle \mathcal{C}[\{d, t\}, f] \mathcal{D} \rangle$$

"When you enter the office you can see a table beneath the window in the left wall. A shelf is adjacent to the rear wall. To its right, in the corner, there is a desk with a chair in front of it."

Assuming we know the size and shape of the room and of the objects to be placed in it (see Figure 6.8), one might expect the semantics of the topological and orientation relations to be sufficient to reconstruct the situation (see Figure 6.9). However, as Figure 6.10 shows, this is not the case. As has been said in previous

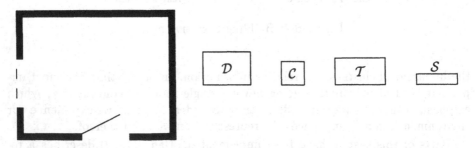

Figure 6.8: Size and shape of prototypical room and objects

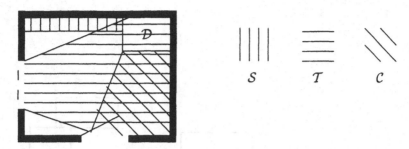

Figure 6.9: Positions according to the semantic of relations

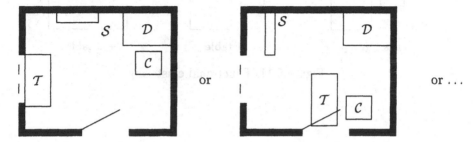

Figure 6.10: Possible positions of objects

chapters, qualitative positional relations are under-determined, allowing many interpretations of the actual position and alignment of objects.

The reason why humans are able to visualize effortlessly is that they use many additional sources of knowledge during the interpretation process leading to a mental or actual sketch of the spatial situation. Some of these knowledge sources, such as basic physical laws, are independent from the specific situation (see section 2.3). Others depend on the particular context. The context determines the sections of commonsense knowledge that are relevant for the given description, e.g., knowledge about a room's typical size and shape, the typical objects and groups of objects in such a room (see Figure 6.11), and the functionality of the objects in the current situation (see Figure 6.12). For example, if the description is given in the context of office layouts, certain default assumptions concerning typical office arrangements are automatically made. Part of the additional constraints required are generated incrementally during the process itself. For example, once a set of objects has been placed, the space they occupy cannot be occupied by other objects at the same time. The hierarchical structure of space also leads to a simplification of the problem by restricting one's efforts to a local conflict in a subspace of the given space. Note that distance is not described by an absolute metric value but is restricted to acceptance ranges. Furthermore, we are not interested in an object's real size but in its relative size. Of course, final placement demands actual object sizes and distances between

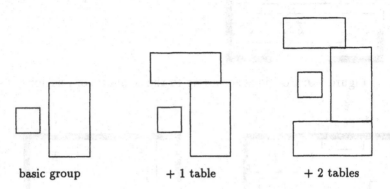

basic group   + 1 table   + 2 tables

Figure 6.11: Functional clustering

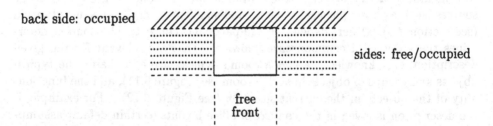

Figure 6.12: Constraints on object's surroundings based on function

objects, but they are established only during the final rendering step after most conflicts have been resolved qualitatively.

The visualization process thus requires several steps (see Figure 6.13):[3]

1. The context established by mentioning a particular type of room (e.g., an office) creates expectations about the prototypical size and shape of the room and objects usually to be found in such as room.

2. Knowledge about functionally related objects (e.g., desk and chair) lead to the creation of functional clusters, which are considered as a unit by the placement procedures. Clustering objects reduces complexity.

3. Objects whose positions are unambiguously fixed (either by the description or by additional knowledge about their function, e.g., a required free front of a desk – Figure 6.12) are placed first.

4. The coarse structure of the prototypical space is used to tentatively distribute the remaining objects and object clusters over the respective subspaces.

5. The structure of these subspaces is examined and conflicts are resolved locally.

6. The final step deals with the actual rendering on the computer screen. This requires selecting quantitative positions, sizes and shapes compatible with the result of the previous qualitative modeling process, and scaling them appropriately for display.

Since we emphasize qualitative aspects, a fluid transition between qualitative information and metrical values is provided. However, this transition can always lead to the need for revisions because of order dependencies. Assumptions not only about the position but also about the shape and size of the objects being represented are necessary, because the relations are interpreted sequentially. All of these might need to be changed when the next relation in the sequence is interpreted. Thus, reconstructive tasks require the maintenance of dependency information and the application of revision mechanisms.

---

[3]The procedures described in this section have been implemented by Daniel Kobler and are further detailed in (Kobler 1992).

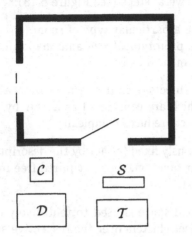

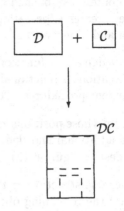

1. Prototypical room and objects

2. Functional clustering

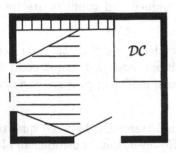

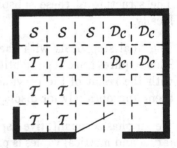

3. Fix unambiguous positions

4. Generate subspaces/assign objects

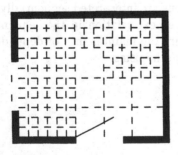

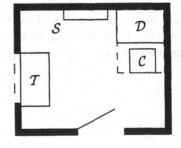

5. Resolve local conflicts

6. Actual rendering

Figure 6.13: Visualization process (example)

Chapter 7

Extensions of the basic model

God in the beginning formed matter in solid, massy, hard, impenetrable, movable particles, of such sizes and figures, and with such other properties, and in such proportion to space, as most conduced to the end for which he formed them.

Sir Isaac Newton, *Optics, 1704.*

In order to explore the full range of reasoning mechanisms and representational alternatives, in this book we have concentrated on a single aspect of spatial knowledge, 2-dimensional positional information. Of course, the qualitative approach is applicable to other spatial aspects as well. In this chapter we will sketch qualitative representations of 3-dimensional positional information as well as of size, shape and distance. We will do so either by reviewing related approaches from the literature or by proposing straightforward extensions of the 2-dimensional framework. These spatial aspects are actually tightly interrelated. Discussing them separately is done for the sake of simplicity and represents an idealization. It requires assuming certain default values for the aspects being ignored. In chapter 4, for example, we assumed well proportioned objects of comparable size and shape.

7.1 3-D space

There is some evidence that the representation of 3-D spatial knowledge might not be as complex as the additional dimension may suggest. Minsky (1986, p. 250), for example, proposes using "direction-nemes" (a square array containing nine regions labeled up, down, left, right, etc.) to represent spatial memories.

This amounts to a reduction of dimensionality from 3-D to layers of 2-D, which could be motivated by the mapping of reality into the two dimensional structure of the visual cortex.

One particularly simple model is to have sets of orthogonal 2-D layers. This idea stems from the observation that in natural language descriptions of 3-D scenes some positions are described based on a 2-D projection on the floor of the scene, whereas other positions are described based on a vertical 2-D slice, but those reference layers are almost never intermixed. That is, the horizontal layer uses the eight orientation relations that we have been using in this book (see section 4.4), while the vertical layer replaces back by above, front by below and the other combined relation accordingly, but does not change the left/right distinction.[1] The distinguished role of the vertical axis due to gravitation (see also section 4.4.1) explains in part the separation of layers: above/below relations often imply physical support of one of the objects by the other and involve concepts such as center of gravity and stability (cf. Funt 1980). If all objects in the focus of attention have a common support, say a desk, then the interesting positional information is in the horizontal plane. The topological relations are not affected by the shift into the 3rd dimension since they are defined on abstract point sets. Some relations (notably overlaps) do, however, have a different physical interpretation as in the 2-D case. Whereas in the case of 2-D projections of 3-D scenes overlapping 2-D objects might correspond to 3-D objects which are above or below each other, in the 3-D world overlapping objects are actually "melt" into one another.

Schwarzer and Högg (1991) propose a different extension of our qualitative model to 3-D based on considering 3 orthogonal projections separately. Each of these projections is reasoned about as in the 2-D case. Then a set of all possible intervals between the end points of the calculated relations is built. Finally all subsumed intervals are eliminated to obtain the tuple describing the derived position. This variant has the advantage of using the unmodified composition tables of the 2-D case. However, it incurs an unnecessary overhead when compared with the layer approach described above, and suffers from the same deficiencies of the straightforward extensions of Allen's temporal calculus to handle 2-D space (see section 8.1).

7.2   Size

As we saw in chapter 4, the relative size of objects plays a role in the determination of both topological and orientation relations. Some topological relations such as included, contains, included-at-border, contains-at-border, equal (collectively called the "containment" relations) can only occur if the objects involved have the appropriate size. For example, only if A is larger than B can A contain B (see section 4.3). Table 7.1 summarizes the size dependencies of topological relations. The relative size of objects also determines the areas of

---

[1]In some 2-D applications such as document layout (see section 8.1) above and below are used as in the vertical layer proposed here.

acceptance for a given orientation relation as explained in section 4.4.4. Being a scalar magnitude, relative size itself can be represented qualitatively by partial orders based on the comparisons $<, >, =$. A distinction has to be made between comparing magnitudes, using the relations just mentioned, and naming sizes in a particular context. This last case is particularly obvious in natural language, where several contextual factors influence the choice of adjectives used to describe sizes (Bräunling 1990). In what follows, we review some approaches from the literature that can be considered to be qualitative. Allen (1983) describes

$\frac{B/C}{A/B}$ toprel	d t o	cOb c	iOb i	=
toprel size	?	>	<	≡
d t o	?		?	
iOb i	<		<	
cOb c	>	>	?	>
=	≡		<	≡

Table 7.1: Size dependencies of topological relations

briefly an extension of his temporal interval reasoner to handle duration, the temporal equivalent of size. The duration relation between two intervals is given by a range (n_1, n_2), that includes the multiplicative factor with which the duration of the first would have to be multiplied to get the duration of the second $(\mathrm{dur}(A) \geq n_1 \mathrm{dur}(B) \wedge \mathrm{dur}(A) \leq n_2 \mathrm{dur}(B))$. Thus, the relation $A - -(0(1)) \to B$ represents the fact that the duration of A is less than the duration of B, since $\mathrm{dur}(A) \geq 0 * \mathrm{dur}(B)$ and $\mathrm{dur}(A) \leq 1 * \mathrm{dur}(B)$ (parentheses about a factor indicate an open endpoint, i.e., eliminate equality). Duration information is encoded in a separate network in which propagation is done by multiplying the respective upper and lower duration limits. Reference durations similar to reference intervals can be used to limit the propagation across scale boundaries.

Mukerjee and Joe (1990), who extend Allen's approach to multi-dimensional spaces (essentially by maintaining tuples of 1-dimensional relations; see section 8.1), base their representation of relative size on the "flush translation operator ϕ". The idea is to observe the relations between two intervals as they move along what the authors call "relation continuum" (i.e., the neighborhood structure). If A [overlaps] B, for example, and the next relation in the flush translation is A [contains-at-border] B, then we know that A is larger than B. They then go on to define an "integer-multiple vector operator" λ, obtained by repeating flush translations equal to an object's size, such that $\binom{2\lambda}{3\lambda} A$ represents a rectangle that is two times as large as A in x, and three times as large as A in y.

Zimmermann (1991) develops a representation for object sizes based on differences and a partial ordering. The relation $A(>, d_1)B$ denotes the fact that "A is higher/larger than B by the amount d_1", since $|A| = |B| + |d_1|$. Negative "sizes" (e.g., depth of a hole) are handled through an index kept outside the lattice relating the absolute values. Questions like "What is the relation between A and C given $A(>, d_1)B$ and $C(>, d_2)B$?" can now be answered, if $d_1(>, d_3)d_2$ is additionally known, since

$$A = B + d_1 = B + (d_2 + d_3) = (B + d_2) + d_3 = C + d_3 \rightarrow A(>, d_3)C$$

The representation is also powerful enough to represent products with natural numbers, e.g., $A(>, B)B$ can be read as "A is two times B", and even size ranges, e.g., "A is between one and two times as high as B", can be expressed as $A(>, d_1)B$ and $B(>, d_2)d_1$.

While representations based on direct comparisons can handle moderately different sizes, other calculi concentrate on differences in the order of magnitude. Order-of-magnitude reasoning is a very common form of qualitative reasoning in engineering by which relations among parameters can be established, for example "A is *much smaller* than B" or "C is of the *same order* as D". Formalizations of this type of reasoning are useful to handle any kind of magnitudes, not just sizes. The FOG system of Raiman (1986), for example, is based on the following primitive relations:

A Ne B: A is *negligible* w.r.t. B
A Vo B: A is *close* to B (and has the same sign as B)
A Co B: A has the *same sign and order of magnitude* as B

It provides reasoning rules to derive new relations, e.g., transitivity rules such as:

(A Vo B) ∧ (B Vo C) → (A Vo C)
(A Ne B) ∧ (B Co C) → (A Ne C)

The O(M) system of Mavrovouniotis and Stephanopoulus (1988) resolves some deficiencies of the FOG system for practical process engineering applications by relating the absolute magnitudes of quantities without reference to their sign. It is based on the following seven primitive relations:

A	≪	B:	A is *much smaller* than B
A	− <	B:	A is *moderately smaller* than B
A	∼<	B:	A is *slightly smaller* than B
A	==	B:	A is *exactly equal* than B
A	>∼	B:	A is *slightly larger* than B
A	> −	B:	A is *moderately larger* than B
A	≫	B:	A is *much larger* than B

Compound relations are disjunctions of two or more consecutive (neighboring) relations. There is a total of 21 compound relations that correspond in part to commonly used concepts such as *less than* $(\{\ll, - <, \sim<\})$ or *approximately equal to* ($\{\sim<, ==, >\sim\}$). O(M) provides well defined semantics for the relations (in two variants, a strict and a heuristic interpretation) and a full-fledged reasoning mechanism to manipulate assignments, constraints, and rules, including an assumption-based truth maintenance system (ATMS).

7.3 Distance

Even though distance cannot be described topologically (see section 4.3) concepts such as "very close", "near", "far away" seem intuitively to be qualitative by definition.

The mathematical concept of distance d between two points v_i and v_j is defined by the three properties:

$$\begin{aligned}
d(v_i, v_j) &= 0 \longleftrightarrow v_i = v_j \\
d(v_i, v_j) &= d(v_j, v_i) \\
d(v_i, v_k) &\leq d(v_i, v_j) + d(v_j, v_k)
\end{aligned}$$

These equations do not necessarily have to be interpreted quantitatively, i.e., by assigning numerical values to the resulting distances. The requirements can be interpreted qualitatively to state that the distance of a point to itself is zero (i.e., the shortest possible), that distance is a symmetric concept, and that the direct distance between two points is shorter or equal to the distances through a third point taken together.

As in the case of sizes, a distinction has to be made between comparing the magnitudes of distances (using the relations $<, =, >$) and naming distances. In the latter case, the types of objects involved and the context in which they are embedded are decisive factors. What it means for object A to be near to object B depends not only on their absolute positions (and the metric distance between them), but also on their relative sizes and shapes, the position of other objects, the frame of reference, and possibly what it takes to go from the position of A to the position of B. What it takes to go from one position to the other might be expressed in terms of metric distance, travel time, effort to be invested, etc. This last point, which also forms the basis for figurative uses of the distance concept, is also the reason why it is so difficult, if not impossible, to represent distance qualitatively out of context. Similar mechanism to those used in the previous section to handle qualitative sizes can be used here, provided the distances can be compared directly. Indeed, once sizes are taken into consideration near and far can be defined relative to the (also relative) sizes of the involved objects. As we saw in section 4.4.4, this distinction is also necessary to correctly define the acceptance areas of orientations whenever objects with extension are involved.

7.4 Shape

Another simplifying assumption made in previous chapters was ignoring the shape of objects. We either considered the delineative rectangle enclosing the objects rather than the objects themselves, or at least assumed them to be convex and without holes (cf. Casati and Varzi ress). In some cases, we even restricted the sides of the delineative rectangle to keep a ration of 1:4 or less. A representational framework for shapes based on qualitative principles is needed to provide high level descriptions of given shapes (recognition task) as well as to derive actual shape designs from functionally motivated descriptions (reconstruction task). In the vision literature there are many, mostly recognition-oriented approaches to the descriptions of forms. Some use uniform volumetric primitives such as generalized cylinders (Binford 1971), others a fixed number of form primitives (Pentland 1986).

As with programming languages, which provide means of describing computational processes (Abelson and Sussman 1985), a language for describing shapes must contain:

- A basic vocabulary of primitive shapes.

- Means of combining simpler shapes to form complex ones.

- Means of abstraction that allow to name and manipulate complex shapes as units.

As part of an approach to the recovery of 3-D volumetric primitives from 2-D images Dickinson (1991) has proposed a model for the qualitative description of shapes. Three basic components are used to generate volumetric primitives (with possible values given in parentheses): cross-section shape (rectangular or elliptical), axis shape (straight or curved), and cross-section sweep (constant or linearly or ellipsoidally increasing or decreasing). Dickinson uses the selection of ten qualitative volumetric primitives from the resulting value combinations shown in Figure 7.1. Complex shapes result from connecting primitives through their distinguished "attachment surfaces". His motivation for qualitative shapes is to provide viewer-centered aspect definitions that are stable under minor changes in scale, dimension or curvature.

Another qualitative model has been developed by Högg (1993) and Schwarzer (1993) as a direct extension of our work. A basic vocabulary of 2-D shapes is defined as a first step towards the description of 3-D shapes. These primitive shapes are classified in *round*, *triangular*, and *rectangular*, and can be varied according to qualitative distinctions of the axial representation, the crossing angle of the axes, and the position of the intersection of the axes. Figure 7.2 shows, for example, all the possible "round" shapes resulting from the variations of the qualitative parameters. 3-D shapes originate from the spanning surface in between a bottom and a top primitive 2-D shape. The relative slope of the surfaces may vary, but for simplicity, a unique correspondance of bottom and top edges and corners is assumed (Figure 7.3a–c). All shapes usually found in natural language descriptions such as *cube*, *cylinder*, etc. can be generated this way. Bottom and top are related by relations such as relative size, position, projection, and height. Curved edges and surfaces can be obtained as modifications (*concave*, *convex*) of the straight shapes (Figure 7.3d–f).

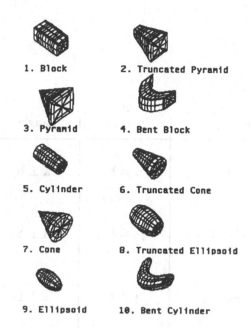

1. Block   2. Truncated Pyramid

3. Pyramid   4. Bent Block

5. Cylinder   6. Truncated Cone

7. Cone   8. Truncated Ellipsoid

9. Ellipsoid   10. Bent Cylinder

Figure 7.1: Ten volumetric primitives (taken from Dickinson 1991)

An auxiliary surface in between bottom and top allows the description of additional shapes (Figure 7.4a,b), including the special cases where bottom and top are points (e.g., a sphere, Figure 7.4c,d).

Identifiable object shapes result from the qualitative proportioning (estimation) of a shape, i.e., from establishing qualitative relations (e.g., $\{<, >, =, \neq\}$) between reference measures, such as length, areas, points, and angles, available in the shape description. This qualitative proportioning contributes to a size and position invariant description. Appendix A contains a summary of the five level shape description vocabulary.

Complex shapes are described by stating the relative position and size of their component shapes, the object consistency (solid, hollow, inverse), and the contact information. The contact information is concentrated in the "contact cover", a description of the contact surfaces of an object (Figure 7.5). The contact cover is often depicted by a planar schema (Figure 7.5), that has the advantage of preserving the neighborhood of the surfaces. There are three types of contacts:

- A simple contact between two surfaces is determined by the three relations *size*, *projection*, and *position* between them.

- A multiple contact is characterized by the 2-D arrangement of its individual contacts.

shape		length	angle	cross
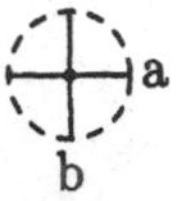	circle	$a = b$	$a \perp b$	$a\ m\ b$
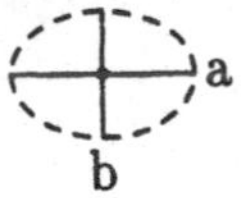	ellipse	$a \neq b$	$a \perp b$	$a\ m\ b$
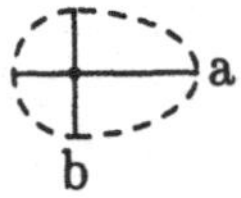	oval	$a \neq b$	$a \perp b$	$a\ ssy\ b$
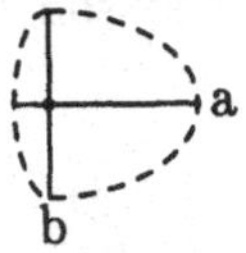	scarab	$a = b$	$a \perp b$	$a\ ssy\ b$
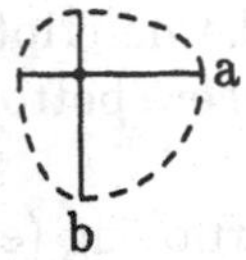	lens	$a \neq b$	$a \perp b$	$a\ s\ b$

Figure 7.2: Parametrization of 2-D primitives

The *length* of the two axes a and b can be the same ($=$) or a be larger than b (*neq*). The axes might intersect at a right *angle* ($\perp$) or not ($\not\perp$).[a] The position of the *intersection* of the axes can be in the middle (*m*), on one side with axial symmetry (*ssy*), on one side without axial symmetry (*s*), or at a boarder (*r*). (taken from Högg 1993)

---

[a] A finer distinction including acute, obtuse, and perhaps straight angles should be more appropriate, in particular for triangular shapes.

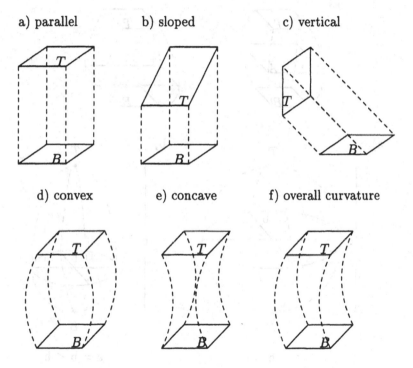

Figure 7.3: 3-D shapes(taken from Högg 1993)

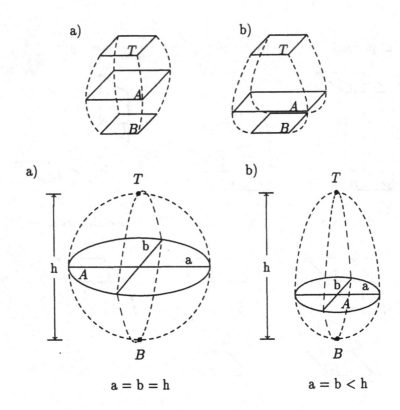

Figure 7.4: Auxiliary Surface(taken from Högg 1993)

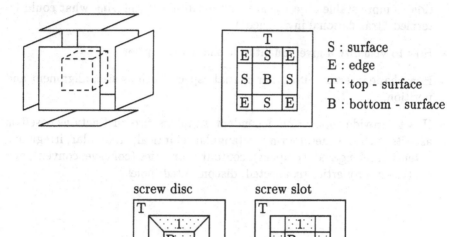

Figure 7.5: Contact cover(taken from Schwarzer 1993)

- Holes and cutouts are described using "inverse" shapes that produce them upon contact with solid shapes.

The description of complex shapes is completed by their "spatial structure" which comprises the relative sizes (proportions) of the components, and their intrinsic orientations. Knowing which surface is intrinsically *up*, for example, determines the orientation of the otherwise undirected contacts.

An important reasoning task about shapes is to determine the similarity of two given shapes based on their qualitative description (disambiguation and matching problems). The matching algorithm must perform several tests: component similarity (number and type of component shapes), contact similarity (number and type), and general spatial structure test. Two shape descriptions are "equal" if they are syntactically identic; they are "equivalent" if the intersection of the sets of shapes that they describe is not empty; they are "similar" if there is only a partial match of features. Various kinds of conversion and resolution rules provide ways of matching equivalent shape descriptions and different descriptions of similar shapes. The current algorithm has several limitations resulting mostly from the fixed order in which the tests are performed. It cannot handle, for example, partial descriptions, i.e., the case where one description is "contained" in the other.

Further issues to be investigated by future research are:

- How to go from viewer-centered descriptions based on immediate perception to more stable object-centered descriptions (achieving what could be termed "translational invariance").

- How to visualize a given qualitative shape description.

- How shape interact with other spatial aspects such as size, alignment and position.

- How to provide application-dependent, simpler sets of distinctions based on aspects such as general form (rectangular, elliptical, triangular, irregular), extension (elongated, compact), contour properties (concave, convex), and continuity properties (connected, disconnected, holes).

Chapter 8

Relevant related work

A tremendous amount of work has been done on the subject of the representation of spatial knowledge covering alternatives that range from precise geometrical descriptions (both analogical and propositional variants), over topological and qualitative representations to linguistically motivated representations.

This chapter surveys the most closely related work, beginning with other qualitative approaches to the representation of space (as well as some papers on time that served as starting points). Section 8.2 reviews then other approaches classified by their most distinctive representational modality (analogical, propositional, etc.), by their principal application (route finding, cognitive maps, etc.), or by their fields of study (linguistics, relational algebras, etc.). For a general overview of recent literature in the area of spatial reasoning see, for example, McDermott (1992) and Topaloglou (1991). Mark et al. (1989) contains references on "Languages of Spatial Relations" and related topics. The core focuses on linguistic studies of how natural languages represent and express objects and relations in geographic space, and on GIS data structures. Bräunling et al. (1990) provide a useful (though a bit dated) general bibliography on spatial reasoning, also available in machine readable form.[1]

8.1 Other qualitative approaches to the representation of space

Even though the qualitative approach has been extensively used for modeling physical phenomena (Bobrow 1984; Weld and de Kleer 1990b), it is only recently that research on qualitative models of space has been undertaken. Allen (1983) introduced an interval-based temporal logic, in which knowledge about time is maintained qualitatively by storing comparative relations between intervals. (Further studies of interval algebras have been done by Nökel (1989) and

---

[1] Hypercard and BibTeX formats. An update is planned as an activity in the proposed HCM Scientific and Technical Cooperation Network "SPACENET: A Network for Qualitative Spatial Reasoning".

Ligozat (1990).) There have been some previous efforts to extend Allen's temporal approach to spatial dimensions (Guesgen 1989; Mukerjee and Joe 1990). However, these extensions just use Cartesian tuples of the one-dimensional relations, loosing the "cognitive plausibility" that Allen's approach has in the temporal domain. Our representation of positions in 2-D space establishes different qualitative relations for the two relevant dimensions topology and orientation. Related research on topological relations in the context of Geographic Information Systems has been done by Egenhofer (1989, 1991), Egenhofer and Al-Taha (1992), Egenhofer and Sharma (1993), and Smith and Park (1992) (this material is discussed in sections 4.3.1 and 5.2.1).

8.1.1   Interval algebras

Allen (1983)   introduced an interval-based temporal logic, in which knowledge about time is maintained qualitatively by storing comparative relations between intervals. That approach has inspired many further developments both in the temporal and the spatial domains. We have already discussed many aspects of Allen's work in this book. Thus, we give here only a brief summary of his model.

Allen takes the notion of a temporal interval as a primitive, as opposed to the point-based approaches prevalent at that time in the AI literature. He establishes 13 mutually exclusive relations describing all possible ways in which two intervals can relate (*before*, *meets*, *overlaps*, *during*, *starts*, *finishes*, and their inverse relations, as well as *equal*). These 13 relations can in turn be expressed in terms of only one primitive relation as shown in a later paper by Allen and Hayes (1985). The relations between a set of intervals are represented in form of a network of constraints. Reference intervals structure the network according to the hierarchical decomposition of the domain. The path finding and constraint propagation algorithms operating on the network use the reference intervals to limit the amount of computation needed (cf. section 5.4.3). In the case of interaction between members of two different branches of a hierarchical decomposition (e.g., parallel subprocesses), this can lead to missing inferences, because the intervals do not share a reference interval. Allen shows how to solve this problem by reorganizing the hierarchy.

Reference intervals are also the basis for a technique to represent the ever changing present moment without having to update most of the relations in the network. Allen also describes a parallel network to record the relative durations of intervals, including mechanisms to express multiplicative factors (cf. section 7.2).

Freksa (1992a)   presents a generalization of Allen's temporal reasoning approach based on semi-intervals (beginnings or endings of events). Although focused on time, this seminal paper is of great relevance to our work, because it introduces the notion of "conceptual neighborhood" of qualitative relations. These neighborhoods, which are motivated by physical constraints on perception, lead to increased inferencing efficiency.

Allen's thirteen relations can be expressed in terms of at most two order relations ($<$, $>$, $=$) between beginnings and endings, because of the transitivity of the order relations and the fact, that the beginning of an event is always before its ending. In the case of incomplete knowledge, using semi-intervals avoids the pitfalls of expressing uncertainty through a disjunction of possible relations. Furthermore, neighborhoods allow to define coarse knowledge, which is given when the uncertainty involves neighboring relations.

The neighborhood definition used by Freksa (1992a, p. 204) is slightly different but equivalent to the one given in section 2.3:

> *Two relations between pairs of events are (conceptual) neighbors, if they can be directly transformed into one another by continuously deforming (i.e., shortening, lengthening, moving) the events (in a topological sense).*

If *events* corresponding to neighboring relations "can be directly transformed into one another", then no intermediate event corresponds to a third relation. Three different neighborhoods are possible in the temporal domain depending on the kind of event "deformations" allowed (Figure 8.1):

A-neighborhood: 3 out of 4 semi-intervals fixed, fourth varies;

B-neighborhood: duration of events fixed, complete events are moved in time;

C-neighborhood: midpoint of events fixed, duration varies.

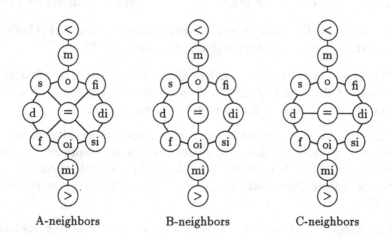

Figure 8.1: Three neighborhoods (Freksa 1992a, Figure 4, p. 207)

Freksa shows that ordering the rows and columns of Allen's composition table according to a useful linearization of the neighborhood structure reveals astounding regularities:

- All resulting compositions are made up of conceptual neighbors. The transitions among compositions are "smooth", i.e., contain no jumps to unrelated relations.

- Only a fraction of the combinatorially possible neighborhoods appears on the table. The occurring neighborhoods can be interpreted as coarse relations (labelled accordingly, e.g., $\{d, f, o_i, m_i\}$ = younger), which can be incorporated into an extended table closed under composition. Furthermore, coarse relations can be used for efficient coarse reasoning and even for fine reasoning with some overhead, but with increased overall inferential power.

- A great deal of symmetry is present and can be used to drastically reduce the size of the table (see section 6 of Freksa 1992a, for details).

8.1.2 Cartesian tuples of relations

Guesgen (1989) introduces "reasoning based on qualitative descriptions of spatial relationships" by defining relations over spatial intervals similar to those defined by Allen over temporal intervals. This is related in spirit to our approach, but the approaches differ in major points:

- Guesgen assumes in his work, that the "exact" quantitative descriptions are as close as we can come to the "real world". We don't.

- While the "cognitive plausibility" of Allen's approach carries over to the one-dimensional spatial case, it gets lost as soon as Cartesian tuples of relations are introduced to handle higher dimensions. Consider for example:

```
O1 <left-of,{attached-to-r, overlapping-r, inside},{left-of,
     attached-to-l, overlapping-l, inside}> O2
```

This is not only hard to read but also not very intuitive! People don't go around decomposing the world in three axes and then determining a qualitative relationship for an interval on each of them.

- Guesgen also suggests using polar coordinates as an alternative. This seems to be more intuitive, and comes closer to our approach, particularly when one forgets about intervals (it saves all the trouble of considering the nearest and furthest point of objects – also something people just don't do!).

- The alleged "shortcoming" of polar coordinates "that the spatial relations may become invalid when the robot changes its position ..." is a problem only as long as the whole system is based only on this form of egocentric representation. However, a transition from an egocentric to a non-egocentric representation framework is a pre-requisite to useful spatial reasoning. In section 5.1 we showed mechanisms for this type of transformation in simple cases.

Mukerjee and Joe (1990) also extend Allen's work on temporal intervals to the spatial domain by using tuples of relations (further aspects of this approach are described in King and Mukerjee 1990; Mukerjee and Bratton 1991). In the one-dimensional spatial case, they consider the relations between an interval and the endpoint of the other one, yielding five regions of interest: + (ahead), f (front), i (interior), b (back), and − (posterior). This implies a reduction of the composition table to 5 * 13 entries, which does not take advantage of as many symmetries as Freksa's compaction does.

In the "multi-dimensional" case (actually only 2-D is shown), they distinguish objects enclosed in boxes oriented with the axes from objects at arbitrary angles. The former case is handled similarly to Guesgen using tuples of relations along the axes, and has the same drawbacks as Guesgen's approach. Nevertheless, Fujihara and Mukerjee (1991) describe an interesting application where these shortcomings are not a problem: the description of document structure. Document layouts deal with rectangular areas aligned with the axes of a rectangular piece of paper. Thus, in this case, considering the relations on each axis separately does not harm (except for documents with an unorthodox embedding of figures in the running text). Blocks containing text or figures can either touch or not (overlap is mentioned as a possibility but not actually used in the examples). The system described (DOCREC) labels the parts of a scanned image as elements of the logical structure of the document, based on the ODA (Office Document Architecture) standard.

The more interesting case of objects at arbitrary angles leads to an intricate framework, in which four directions and four quadrants provide the basis for a two part relation including the relative direction (expressed in terms of quadrants) and the relative position. The representation of relative direction is grounded on intrinsic fronts, based on the assumption that most objects have a distinguished front. As we discussed in section 4.4.3, this is not the case. The representation of relative positions is based on projecting the boundaries of the object in the direction of the intrinsic front to obtain its "lines of travel". The lines of travel of two objects form a "collision parallelogram", w.r.t. which the relations *behind*, *after*, *inside* (corresponding to qualitative areas), and *back*, *front* (qualitative points) are defined. The relative position of two objects is specified by the quadrant information (dir(A/B)), and the positional relations pos(A/B) and pos(B/A). These complex relations lead to huge composition tables involving 676 * 676 = 456976 entries for the 2-D case![2] Even using various redundancies and symmetries, the size of the table can be reduced only by a factor of eight. Another aspect of Mukerjee and Joe's work is their use of several operators such as flush-translation, reflection, and integer vector multiplication to handle relative size and discrete rotations. For example, the flush-translation operator ϕ moves A until it flush with B, allowing us to determine whether A is longer, equal, or shorter than B by establishing the relation between ϕA and B. This framework has been applied to qualitative vision (Mukerjee 1991a) and

---

[2]676 = 13 * 13 * 4, result from two times the 13 position relations, and the four direction quadrants.

qualitative geometric design (Mukerjee 1991b) and more recently extended to model 3-D shapes.

8.1.3   Other relational approaches

Freeman (1975), an early paper on *The Modeling of Spatial Relations*, is itself mostly a review of previous work particularly from psychology. It is interesting mainly because, on the one hand, it shows the kind of ad hoc approach that characterized many AI-systems in the 70's, on the other hand, it already pointed at many of the essential "ingredients" of a qualitative model of space. Freeman distinguishes two types of relations between objects: those involving comparisons of the properties of objects (e.g., *larger, darker, smoother*), and those involving their relative position. He concentrates on the last class consisting of the 13 relations: *left of, right of, beside, above, below, behind, in front of, near, far, touching, between, inside, outside*. He then lists some properties of that very heterogeneous group of relations such as arity, reflexivity, symmetry, antisymmetry, transitivity, and inversion. Of special interest in relation to acceptance areas (cf. section 4.4.4) is his discussion of the semantic content of the *left* relation (in the 1-D spatial case). While a fuzzy set approach (Zadeh 1965) allows the specification of the "degree" of leftness by using a continuous characteristic function in the range [0, 1], the approach of Winston (1975) uses the center position of one of the objects as threshold in the comparison with the vertices of the other object. An extension of the latter method would allow two thresholds, with an indeterminate zone in between.

Although with respect to a different issue, the context sensitivity of spatial relations was already recognized in the early work on the subject. The largest part of the paper reviews Clark and Chase's (1972) psychological experiments on relational encodings. Assuming that humans encode spatial knowledge as an interpretation (i.e., named objects and their relations) rather than as a perceptual entity, they establish empirically some interesting rules:

1. People normally code pictures in positive terms (i.e., descriptions are given in terms of what *is* there rather than what is not there).

2. People refer to the locations of objects positively, where upward and forward from the observer are the positive directions.

3. Perceptually prominent objects are used as reference objects in any relation involving them.

4. Perceptual comparisons are also preferentially coded in positive form on the underlying dimension.

The criteria for preferring a particular dimension over another one are also reviewed. This has been the subject of many subsequent studies, e.g., Lang and Carstensen (1989).

Pfefferkorn (1975) describes a design system for furniture layouts that incorporates topological and metric constraints. The internal representation of space is a numeric one, describing objects and floor plans in terms of convex polygons consisting of sides and points. Design requirements, however, can be expressed in form of qualitative constraints, e.g., "the table should be to the left of the lamp", "the desk should face the window", "TV should be visible from couch". The system, called DPS, is capable of selecting positions for irregularly shaped objects such that all design constraints are satisfied. It employs a heuristic search technique using a tree structure to remember the different alternatives created. To minimize the number of positions examined, DPS tries first to place objects in various directions in free corners of the space available, checking if all relevant constraints are satisfied. To assist in finding these positions, spatial operators exist for creating lists of contiguous space blocks, boundaries of contiguous space blocks, and corners of boundaries. These operations use explicitly stored adjacency information. The "diagnostic-search" strategy employed, moves forward in a depth-first manner until a difficulty is encountered. Then, instead of backtracking, remedial actions are undertaken, including different order placements. A planning phase done prior to the problem solving phase gives the system a "sense of direction" that further reduces the number of placements searched.

In spite of these heuristics, the system is described as slow by its author. One of the reasons is, that planning and placement procedures operate directly on the quantitative internal representation. Given that the design constraints are qualitative, much of the planning and initial placement could be done at the qualitative level with less computational overhead. Furthermore, most of the structural properties of space have to be explicitly expressed (and maintained, whenever something changes), for example, in form of adjacency lists.

Di Manzo, Giunchiglia, and Pino (1985) discuss a system, called NALIG, for generating images of a scene from a natural language description. The system makes extensive use of qualitative reasoning techniques, even though the final positioning step must produce numerical coordinates. During the inference process, various sources of knowledge are taken into consideration including the semantic of spatial relations, the geometrical and physical properties of the objects involved, naive representations of common physical processes, prototypical arrangements, object functionality, and so on. The influence of all these factors is demonstrated through an analysis of the conceptualization of the H_CONTACT relation. Using the prototypical shapes and dimensions of the objects involved, the various possible positions and orientations are checked against equilibrium conditions, functional considerations, and typical configurations. The internal space occupancy is stored using oct-trees for efficiency. While the approach stands out for taking the many sources of world knowledge into consideration, the scope of that knowledge is not clear. Furthermore, no well-founded linguistic theory stands behind the semantics attributed to the natural language statements.

Green (1987)   describes another system centered on design called SPACES. Her emphasis lies on incorporating the kind of commonsense spatial knowledge that is used in the early stages of design, where constraints are expressed in non-numerical terms. She proposes a calculus based on a set of primitive spatial entities such as spaces, boundaries, and points. These entities have associated attributes: Spaces have boundaries and areas, boundaries have length and orientation, and so on. A set of construction operators allows the designer to define spatial arrangements in terms of the primitive entities. Examples of these operators are: Create a space or boundary, divide a space, place or move a space or boundary, join two spaces, remove a boundary. The PROLOG implementation of the system assumes rectangular spaces, which allow unambiguous definitions of the relative positional relations above, below, left-of, and right-of (only one of which holds at a time). This is certainly a very strong limitation on the usability of the system. Besides the definitions for entities, attributes, and relations, and a set of predicates acting as operators, the system maintains a database of statements that form the internal model of the spatial arrangement. The author admits that such an internal model, containing relations expressing local views, makes it difficult to maintain spatial consistency: The space affected by the placement of a new division has to be *searched* for in the model. She suggest the use of analogical representations to complement her system.

Forbus, Nielsen, and Faltings (1991)   present a framework for qualitative kinematics (geometric aspects of representing and reasoning about motion) based on combining metric diagrams with a qualitative place vocabulary. They claim that there is no powerful, general purpose, purely qualitative spatial representation (poverty conjecture), and thus introduce the place vocabulary as a symbolic description of shape and space, grounded in metric diagrams. The basic inferences needed in qualitative kinematics are finding potential connectivity, kinematic states, mechanical states, and state transitions. Although it is true that there are no general-purpose representations, whether qualitative or not, this alone is no reason to restore to metric information or to claim that qualitative spatial representations in the spirit of Allen's temporal logic are virtually useless. Several of the models described in the previous sections and the one developed in this book prove this claim wrong.

Freksa (1991)   discusses in general terms many of the basic properties of qualitative representations on which our work is based, and we have referred to this paper extensively, in particular in chapter 2. He points out the unique role of spatial knowledge for cognitive systems resulting from the fact that it can be perceived through multiple channels. This makes it useful even for non-spatial domains: spatial metaphors often help us to understand complex matters. He makes a careful study of the role of abstraction in spatial reasoning by stating which properties of space we want to abstract from and which ones we want to keep. Among the latter are uniqueness constraints, topological properties and the conceptual neighborhood structure. He envisions a "spatial inference

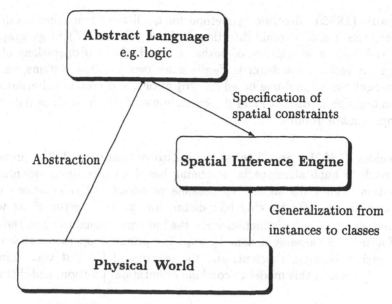

Figure 8.2: Spatial inference engine (Freksa 1991, Figure 1, p. 364)

engine" providing just the right level of abstraction as shown in Figure 8.2. He gives a general characterization of qualitative knowledge and relates it to "mental images" using the aquarium metaphor as an example.

Cohn, Cui, and Randell (1992) summarize a theory of space and time based on a calculus of individuals founded on "connection" and expressed in the many sorted logic LLAMA. The basic set of dyadic topological relations defined using the primitive $C(x,y) =$ 'x connects with y', x and y being regions is (Randell, Cui, and Cohn 1992d): $DC(x,y)$ ('x is disconnected from y'), $P(x,y)$ ('x is a part of y'), $PP(x,y)$ ('x is a proper part of y'), x=y ('x is identical with y'), $O(x,y)$ ('x overlaps y'), $DR(x,y)$ ('x is discrete from y'), $PO(x,y)$ ('x partially overlaps y'), $EC(x,y)$ ('x is externally connected with y'), $TPP(x,y)$ ('x is a tangential proper part of y'), $NTPP(x,y)$ ('x is a nontangential proper part of y'). The subset of mutually exhaustive and pairwise disjoint relations $\{DC, EC, PO, =, TPP, NTPP, TPP^{-1}, NTPP^{-1}\}$ corresponds exactly to the set of eight topological relations derived in section 4.3.1. Furthermore, sum, intersection, and complement of regions are defined as functions, as well as the primitive function conv(x) for "the convex-hull of x". conv is used to define the relations $INSIDE(x,y)$ (x is inside y), $P\text{-}INSIDE(x,y)$ (x is partially inside y), $OUTSIDE(x,y)$ (x is outside y), and their inverses. This framework has been used to perform qualitative spatial simulation (Cui, Cohn, and Randell 1992), and to reason about physical systems such as a force pump (Randell, Cohn, and Cui 1992c).

Frank (1992)   describes a method for qualitative reasoning about distances (*far*, *close*) and cardinal directions[3] (*N*, *E*, *S*, and *W*) in geographic space. It is based on an algebra of paths on which the two operations of *inversion* and *composition* are defined. He discusses two direction systems, one based on triangular areas and one based on projections, and presents alternatives for the combination of distance and direction, some of which produce only 'Euclidean approximate' results.

Freksa (1992b) and Freksa and Zimmermann (1992)   present an approach to qualitative spatial reasoning based on directional orientation information. They distinguish 15 possible positions and orientations of a point based on the left/straight/right distinction w.r.t. a vector $\vec{ab}$ as well as the front/neutral/back distinction w.r.t. the lines orthogonal to $\vec{ab}$ on the end points of a and b. Reasoning is done by applying primitive operators such as identity, inversion, homing, shortcut, and the inverses of the last two. Zimmermann (1993) extends this model to combine orientation, position, and distance.

Fuhr, Kummert, Posch, and Sagerer (1992)   use the neighborhood of qualitative relations to predict changes while interpreting image sequences. Based on a continuity assumption, qualitative derivatives, and the order of qualitative changes, their model is able to control the focus of the expectation-driven object recognition processes.

Clementini, Di Felice, and van Oosterom (1993)   extend topological relations to include not only the area/area case previously studied but also the line/area, point/area, line/line, point/line, and point/point cases. Furthermore, instead of just distinguishing between empty and non-empty, the dimension of the intersection (0D=point, 1D=line, 2D=area) is taken into account. Eliminating impossible cases this yields 52 cases, which are still far too many for user interaction. Thus, an "object-calculus" involving only five overloaded topological relations (*touch*, *in*, *cross*, *overlap*, *disjoint*) is proposed. The authors show these relations to be mutually exclusive and to cover all cases previously studied.

Freksa and Röhrig (1993)   make a comparative study of nine research projects on representation of spatial knowledge. Among the dimensions used to characterize the various approaches are:

- **Frame of reference**, which can be of various types (Cartesian, polar, etc.), have either a local or a global scope, and differ in its alignment (internal, external).

- **Representational primitives**: Whether to use abstract points or spatially extended objects as the basis of representation.

---

[3]Cardinal directions are easier to analyze then relative orientations because the frame of reference is fixed in space.

- **Spatial aspects represented:**[4] Topology, arrangement, distance, orientation, and which ones are actually represented qualitatively.

- **Dimensionality** and **modularity** of the representation.

- **Granularity** (i.e., fine vs. coarse, mixed).

- **Vagueness**, i.e., their ability to handle incomplete and imprecise knowledge.

Given the different contexts in which the nine approaches analyzed have been developed (ranging from formal, logic-based models over database oriented research and GIS applications to cognitive motivations), it is very difficult to make a clear cut classification. On the other hand, as the authors point out, there is a remarkable degree of convergence on the issue of qualitativeness.

Röhrig (1993) continues this investigation and shows how the different qualitative approaches can be modeled using an uniform one-dimensional representation based on a cyclic binary order relation called CYCORD. In particular, it is shown how to perform inferences based only on the transitivity property of the CYCORD relation.

Ligozat (1993a) also examines several qualitative models for spatial reasoning and proposes similar parameters as Freksa and Röhrig (1993) to distinguish between them. He reconsiders the one-dimensional case, and both global (i.e., based on projections on axes) and relative 2-D cases. The global 2-D calculi include those based on the $<, =, >$ relations on the axes, and those based on the full set of Allen's interval relations. In (Ligozat 1993b) he provides a systematic way of deriving calculi based on qualitative triangulation such as Freksa and Zimmermann (1992), which allow inferences about the relation between points in 2D.

Vieu (1993) presents another logical framework for reasoning about space based on mereology (an axiomatization of part-whole relations), which includes formalizations of topology, distance, and orientation. Vieu's model is also an extension of Clarke (1981, 1985) on which Cohn's et al. work above is based. This theory takes "individuals" or "bodies" as primitive entities instead of points, and uses a single "connection" relation. Two individuals are "connected" when they share some part or when they are in contact. Points are defined as sets of individuals. Besides the classical mereological relations "part of", "proper part of", and "overlaps", connection allows the definition of "external connection", "tangential part", and "non-tangential part".

---

[4]Missing from this list are size and shape, which can also be represented qualitatively (see sections 7.2 and 7.4 respectively).

8.2   Other approaches to the representation of spatial knowledge

8.2.1   Representation modalities

As a result of the extended "imagery" debate in cognitive science (Kosslyn et al. 1981; Pylyshyn 1981) approaches to the representation of spatial knowledge tend to fall into one of the categories "propositional" or "pictorial". There are several problems with these two extreme positions:

The propositional approaches focus primarily on formal properties of the representation such as soundness and completeness (Reiter and Mackworth 1989). While doing so, however, they are forced to explicitly express the rich structural properties of space using propositions. Thus, basic structural properties of physical space such as those mentioned in section 2.3 must be stated explicitly leading to an enormous computational overhead even for moderately realistic applications.

On the other hand, pictorial representations (Funt 1980; Khenkhar 1989), while preserving many of the spatial properties mentioned above, model some aspects of space on too low a level. For example, by defining a fixed grid of a given granularity (each field of which can be either filled or not) to represent occupancy, an arbitrary discretization of the domain is introduced. This may lead to wrong results, because binary decisions are forced at an inadequate stage of processing.

However, the desirable properties of both propositional and pictorial representations can be combined. This has been recognized by several authors (Pribbenow 1990; Chandrasekaran and Narayanan 1990), who nevertheless propose hybrid models that combine them by interfacing them as *separate* representations. The qualitative representation of space we propose in this book *integrates* them by exploiting the generalization capabilities of abstract relational representations, while inherently reflecting domain constraints such as the neighborhood of the positional relations.

This is possible, because pictorial representations are just a special case of the more general class of analogical representations (Sloman 1971), i.e., those, where important properties of the represented domain are intrinsically represented by the inherent properties of the representing domain. The structural similarity between the represented and the representing domains doesn't need to be an isomorphism and can be given at an abstract level.

In what follows we discuss exemplarily one approach from each of the modalities depictorial and propositional, as well as an hybrid system.

Depictorial representations

Several authors have proposed depictorial approaches, including Adorni et al. (1983), Habel (1989), Khenkhar (1989), Haugeland (1987), Lindsay (1988), Mohnhaupt (1987), and Sander (1991). We comment on one of the earliest systems with an implementation.

Waltz and Boggess (1979) were among the first to actually use a depictorial representation to interpret natural language statements about the physical world. Their model builds an internal "visual analog" representation of the spatial description given as input. This internal model is then used to answer questions about the particular configuration by "inspection". They contrast this approach with a deductive approach in which similar queries might easily lead to a combinatorial explosion.

The system is based on detailed definitions of prototypical objects (containing shape, dimension, weight, etc.), and LISP functions defining spatial prepositions. Input to the program are either statements (naming or locative ones) or questions. Output is either a set of coordinates for each object in the "mental model" or an answer to a question. The functional definitions of prepositions are based on primitives such as *contiguous*, *supported*, *interior*, etc.

One of the drawbacks of maintaining an internal 3-D coordinate representation is that the objects have to be placed at a definite spot on interpretation of the locative statements, even if they contain ambiguities. Thus, inferences not implied by the corresponding statements might be drawn. The difficulty of representing uncertainty is a common defect of all depictorial approaches.

Propositional representations

Many approaches, including the relational approaches discussed in section 8.1.3, can be considered to be propositional. In this subsection we want to discuss the work of Reiter and Mackworth (1989) as an extreme case of a complete logical reconstruction of a spatial reasoning system (see also Schwartz 1989; Bäckström 1990).

Reiter and Mackworth (1989) propose a logical framework for depiction and image interpretation. As an example, they provide a logical reconstruction of Mackworth's Mapsee system designed to interpret hand drawn sketch maps of geographical regions. The interpretation of an image is defined to be a model in the strict logical sense. Thus, all aspects of the spatial domain under study must be formalized in form of axioms. Three sets of axioms are required in this framework: image axioms, scene axioms, and depiction axioms. We list here a few examples of each class of axioms to give a flavor of this type of logical specification.

In the simplified version of the Mapsee system used as an example there are only two types of image objects: chains and regions.

$$(\forall x)\text{image-object}(x) \equiv \text{chain}(x) \vee \text{region}(x)$$

$$(\forall x)\neg(\text{chain}(x) \wedge \text{region}(x))$$

The following relations may hold between image primitives: T-junction(c,c), χ-junction(c,c), bounds(c,r), closed(c), interior(c,r), exterior(c,r), under closed world and unique names assumptions, and coherence requirement for constants occurring in the closure axioms. The scene domain axioms specify a taxonomy of scene objects (linear objects like road, river, shore; areas like land and water) and general facts about them such as for example:

$$(\forall x, y)\text{RIVER}(x) \wedge \text{RIVER}(y) \supset \neg\text{CROSS}(x, y)$$

(Rivers do not cross each other.)

$$(\forall x)\text{SHORE}(x) \supset \text{LOOP}(x)$$

(Shorelines form closed loops.)

Type constraint axioms such as

$$(\forall x, y)\text{CROSS}(x, y) \supset \text{SCENE-OBJECT}(x) \wedge \text{SCENE-OBJECT}(y)$$

are also required.

Next, the mappings between image and scene domains must be specified. These are restrictions such as that every scene object is depicted by an unique image object:

$$(\forall s)\text{SCENE-OBJECT}(s) \supset (\exists! i)\text{image-object}(i) \wedge \Delta(i, s)$$

which precludes occluded objects in the scene. ($\Delta(i, s)$ is a predicate meaning that image object i depicts scene object s.) There is also a need for taxonomic mappings (regions and chains in the image depict areas and linear objects in the scene, respectively) and relational mappings, establishing for example, that tee relations in the image depict join relations in the scene, and vice versa:

$$(\forall i_1, i_2, s_1, s_2)\Delta(i_1, s_1) \wedge \Delta(i_2, s_2) \supset tee(i_1, i_2) \equiv \text{JOINS}(s_1, s_2)$$

Once having specified these three sets of axioms, the interpretation process can be done as model construction. First, however, a set of simplified axioms is derived by applying some formal results. The complexity of this process poses some serious problems. In general, it is undecidable whether a set of formulas has a model, and there may be infinitely many models. The inflexibility of this approach is evident in the following quote (Reiter and Mackworth 1989, p. 135):

> "Is there anything special about vision which precludes this problems?
> [...] The most promising observation is that an image is finite. [...]
> Whenever this is the case, quantified formulas reduce to propositional
> ones and image interpretations are all computable."

As we have argued in this book, there is much more that is "special" about images than being finite. The constraints of physical space, which through their partial formalization in the logic-based approach contribute to the complexity of the model construction process, actually place tight restrictions an what makes sense on an image. In a sense, the formalization of space abstracts from its rich structure without fully re-capturing that structure in form of explicit constraints (cf. Figure 8.2).

Hybrid approaches

Chandrasekaran and Narayanan (1990) describe an hybrid system on which we comment below. Habel and Pribbenow (1989) and others in the LILOG-project have also developed an hybrid system for computing the meaning of localization expressions involving spatial prepositions that we describe later in the context of linguistically motivated work. For general issues involved in the development of hybrid systems see Aiello and Nardi (1991).

Chandrasekaran and Narayanan (1990) propose an hybrid system intended to integrate "modality-specific" (e.g., visual) mechanisms with abstract symbolic representations. The "Image Representational System" (IRS) is made out of "Image Symbol Structures" (ISS). ISS are hierarchical discrete symbol representations consisting of parameterized primitives, which encode only the intrinsically visual aspects of a scene. Among those primitives are the "symbolic percepts" (S-percepts) containing descriptions of spatial relations among the volumetric parts of an object. The corresponding "analogic percepts" (A-percepts) are the mental images that result from the interpretation of the S-percepts. The authors sketch how to apply this representation to kinematic simulation, i.e., predicting the motion of rigid objects, using a multi-level description in which all but the lowest levels are symbolic, and the lowest level is "imaginal" (depictorial). The description of the reasoning at both the symbolic and depictorial levels is too general as to be able to judge if and how it would differ from previous work by Funt (1980), Gardin and Meltzer (1989), or Steels (1990).

8.2.2 Cognitive maps and route finding

Many studies on the representation of spatial knowledge have been done with applications such as the assimilation of cognitive maps and route finding in mind. We describe here some of the classical work in the area. Reviews of the development of spatial cognition are given by Hart and Moore (1973), Cohen (1983), Stiles-Davis, Kritchesvsky, and Bellugi (1988), Blades (1991b, 1991a), Tversky (1991). For studies on the cognitive representation of spatial relations see Stevens and Coupe (1978), Baird (1979), Baird, Merrill, and Tannenbaum (1979), Mani and Johnson-Laird (1982), Olson and Bialystok (1983), McNamara (1986), Bialystok and Olson (1987).

Kuipers and Levitt (1988) describe a series of influential systems for navigation and mapping in large-scale space, i.e., space that cannot be perceived at a glance. Navigation in a large-scale environment requires building a cognitive map of the environment by integrating many individual observations. The authors propose a four-level semantic hierarchy of descriptions: sensorimotor (I/O relations with environment), procedural (primitive procedures for accomplishing place-finding and route-following tasks), topological (description in terms of places, paths, landmarks, and regions, linked by topological relations), and metric (description in terms of same entities as before, but linked by relations such as relative or absolute distances and angles). This hierarchy has been applied to the design of three programs briefly described in what follows:

The TOUR model (Kuipers 1978) assumes a sensorimotor world consisting of an alternating sequence of views and actions, and procedural behaviors given in terms of production-like schemata (goal, situation, action, result). It has two levels of topological information: a network of places and paths, and the containment and boundary relations of places and paths with regions. The metric information can involve either the local geometry at places along paths, or local orientation frames w.r.t. a global frame of reference.

Whereas the TOUR model was conceived for network-like environments such as urban networks of places and streets, the QUALNAV model (Levitt and Lawton 1990) handles open terrain with visible landmarks. QUALNAV includes a coordinate-free, topological representation of relative spatial location, and integrates metric knowledge of relative or absolute angles and distances. The qualitative representation is based on drawing imaginary lines between landmarks such that the position of the robot can be determined by observing the order in which the landmarks are seen (see also section 4.2). This representation is at the heart of a qualitative navigation simulator. The NX robot (Kuipers and Byun 1987) is a simulated robot that extends the TOUR model by handling continuous sensory input and continuous motion.

McDermott and Davis (1984) develop a model for the representation of spatial knowledge in the context of a route planning system called SPAM. Their representation has two components: a propositional component to express topological facts, implemented as an indexed set of assertions, and a "fuzzy map" to store metric facts in form of a map. A map is a table containing for each object relevant attributes such as position, orientation, height, etc. Uncertainty is handled by storing ranges (called 'fuzzes') instead of single values for every attribute, and by allowing multiple references to an object in different frames. By contrast, our approach uses under-determinate relations that avoid the computational overhead of range calculations. Retrieval is done through theorem proving in the assertional database and through linear programming methods—to solve queries expressed in form of numerical optimization problems—in the fuzzy map. Assimilation is done either through "fuzz constriction", where a new fact is used to tighten the fuzzes of known quantities, or through re-mapping, when it is necessary to rearrange the frames of reference or create new ones. In order to represent the fact that objects are only fuzzily located, "fuzzboxes" are drawn within which the object is believed to lie. These fuzzboxes have sharp edges (relativized by the fact that probabilistic algorithms are used). The authors argue that a decision maker must eventually use arbitrary criteria on borderline cases. As we have suggested in this book, another way to go is to keep relative information (i.e., avoid taking the decision), and to decide only when needed in the actual context.

The MERCATOR system (Davis 1983, 1986) is a successor of SPAM in which cognitive maps are constructed from scene descriptions. Objects, i.e., their shapes, are represented as polygons, boundaries are represented as directed edges. Local dimensions are recorded in terms of ranges allowing various grain sizes. A given object may have several region descriptions. Objects are organized hierarchically by containment.

Yeap (1988)   The "Computational Theory of Cognitive Maps" of Yeap (1988) emphasizes the acquisition of a "raw cognitive map" from perceptual information (reduced for simplicity to a "$1\frac{1}{2}$-D Sketch" containing the projection to two dimensions of all relevant surfaces in a room). Although taking into consideration many cognitively plausible elements, such as a short term restricted representation of the immediate surroundings, the algorithms work with exact quantities for sizes and angles of perceived surfaces. In section 6.1 we show how such an approach can benefit from the introduction of a qualitative representation.

8.2.3   Linguistically motivated research

We have been using natural language spatial descriptions as a way to *access* cognitive spatial concepts. Linguistic phenomena occurring during verbal descriptions of space are a subject of study on its own right. Many studies concentrate on spatial prepositions (Bennett 1975; Hill 1982; Levelt 1986; Bierwisch 1988; Herweg 1989; Hays 1990; Pribbenow 1990, 1991), because they are usually at the core of statements conveying spatial information (the other major source being verbs of motion). Spatial prepositions are often classified in topological, directional (or projective), and path prepositions. The issues related to the interpretation of projective prepositions were discussed in the context of the frames of reference in section 4.4.3 (see also Retz-Schmidt 1988). Determining the meaning of spatial prepositions is not easy, because a single preposition (e.g., "in") may have many context-dependent meanings. For other linguistic aspects see also Habel (1988a, 1988c, 1988b, 1989), and Lang (1987).

Herskovits (1986)   postulates in her study of English spatial prepositions a prototypical "ideal meaning" that changes according to various use types. This ideal meaning refers to the geometric descriptions of the objects involved. The use types can require a "sense shift" or just a "tolerance shift" in which the ideal meaning is modified only gradually. The choice of transformation from the ideal meaning is guided by the four principles saliency, relevance, tolerance, and typicality. The type of qualitative model of spatial knowledge developed in this book could serve as a mechanism to represent and reason about the ideal meaning of spatial prepositions by expressing them in terms of primitive topological and orientation relations.

Habel and Pribbenow (1989)   and others involved in the LILOG-project, developed an hybrid system for computing the meaning of localization expressions involving spatial prepositions. It uses a propositional and a depictorial component that complement each other. An explicit axiomatization of the topological and geometrical theory required to describe the spatial concepts of localization is difficult to get, because of the very richness of spatial knowledge. For the same reason it is likely to be intractable. Thus, the propositional component handles the language-specific processing and the general properties of spatial concepts. The depictorial component is triggered by special rules of the propositional component. It provides means for establishing the conceptual meaning of a localization fact by determining the corresponding localization area. The depictorial component computes the concrete size and position of localization areas, and is thus able to answer complex spatial questions. Depictions are implemented using cell matrices (finite 2-D arrays), on which two kinds of processes operate: Imagination processes, which construct images based on assumptions about typical form of object classes, and inspection processes, which examine all matrices to answer spatial queries (for example, by determining if there is a barrier interrupting the topological connection between reference object and primary object, in which case they cannot be "near").

8.2.4 Relational algebras

Formal work on relational algebras to represent spatial knowledge has been done mostly in the context of database research. From the point of view of database theory, spatial problem solvers consist of three fundamental components: representation, reasoning, and management. It is this last component, the management of large spatial databases, that is of special interest in this area of research. By "management" it is meant how to organize the spatial (and maybe even related non-spatial) information in a way that efficient storage and retrieval becomes possible. An overview of management issues for large spatial knowledge bases is given by Topaloglou (1991).

Güting (1988) proposes a geo-relational algebra as a model to incorporate geometric information in a relational database system. This is done by introducing new geometric data types (point, line, area), special predicates ($=$, $\neq$, inside, outside, intersects, isNeighborOf), and operators. The operators can be classified in transformers, e.g. intersection: line $*$ line $\rightarrow$ point (other transformers are overlay, voronoi, closest), those returning atomic objects (e.g., convexHull: point$* \rightarrow$polygon), and those returning numeric values (e.g., dist: point $*$ point $\rightarrow$ num). The geo-relational algebra is intended to be both a user model (interface) and the basis for the implementation of a database management system.

Laurini and Milleret (1988) assume a less sophisticated relational database model. They concentrate on showing how objects (being here closed, non-overlapping polygons) can be stored in a relational database, and how three particular tasks can be solved in the different models. The spatial models analyzed are a conventional wire-frame model requiring computational geometry algorithms for query evaluation, the Peano relation model allowing the use of an algebra, and a mixed model requiring both geometry and algebra. The three tasks they study are:

- Point-in-polygon queries: the answer expected is the name of the polygons in which the point lies.

- Region queries: the answer expected is the names of all polygons that include or intersect a given polygon.

- Vacant places: resulting in the parts of a given polygon that do not intersect with other polygons.

The tradeoffs shown by the application of the three models to these three tasks indicate that the intended use and scope of the spatial data must be taken into consideration in the design of spatial database systems.

Chang, Shi, and Yan (1987) introduced the symbolic projection schema in the context of pictorial databases. Two dimensional spatial arrangements are represented by projecting them into two "2D-strings" along the vertical and horizontal axes. If several objects overlap in the projection, then the two operators "<" (for right/left or bottom/up) or "=" (for same position) indicate the order in which they do so. Whenever the overlap constellation changes a new element of the string is created. For example, the string $A = D < B < C, A < B = C < D$ represents a scene in which D is above A, and B and C are to the right of A and in between D and A. The 2D-string serves as an iconic index for the picture, where retrieval corresponds to a 2D-string subsequence matching. Various extensions of this model have been proposed including further operators and a combination with quad-trees (Chang and Li 1988), local operators to handle overlapping objects (Jungert 1988) (object partitioned into subparts at bounding lines of overlapping objects). Finally Lee and Hsu (1991), in an effort to provide a more efficient "cutting" mechanism came up with a set of operators that is very similar to Allen's 13 relations. While in the context of pictorial databases (where the actual spatial content of the pictures is not relevant) 2D-strings are useful as indexing mechanism, as a representation of spatial knowledge they suffer from the same drawbacks of Guesgen's approach. Recently, further extensions has been done to take the observer's point of view into account (Jungert 1992), to exploit rotation invariance (Jungert 1993a), and to handle object shapes for qualitative matching (Jungert 1993b).

Papadias and Sellis (1992, 1993) proposed a similar approach to symbolic projections. They use 2D array structures called spatial indexes to represent a set of spatial relations between representative points. These points are placed in the proper cells in the spatial index and are thus different from the projections used above. Spatial indexes can also be used to represent direction relations using either one or four points per object, and topological relations by analyzing the indexes stored in the cells.

Chapter 9

Conclusion

Vagueness is a solution rather than a problem.
P.N. Johnson-Laird, *Mental Models, 1983.*

In this final chapter we give a summary of the book and highlight its main contributions. It also contains a general discussion of current limitations and future research issues, including ways in which qualitative positional knowledge can be integrated with other types of spatial knowledge (e.g., path knowledge, metric knowledge) and non-spatial knowledge, and its relationship to linguistic concepts and perceptual representations.

9.1   Main contributions

As main contributions of this book, we would like to highlight the following:

- Having found cognitive space to be qualitative in nature, we propose using qualitative representations, which are compact, capture relative information, are under-determined, handle vague knowledge, are independent of fixed granularities, have varying information content, and benefit from mutual constraints.

- Not only cognitive but also computational factors suggest a qualitative approach to the representation of spatial knowledge. When faced with incomplete or uncertain information, quantitative approaches have to deal with ranges of possible values and cumbersome algorithms. We show, that the inherent under-determination of qualitative representations "absorbs" the vagueness of our knowledge, reducing the complexity of the algorithms involved.

- We demonstrate the usefulness of the qualitative approach by providing a model for the qualitative representation of positional information in 2-D space based on topological and orientation relations.

- We introduce "abstract maps", containing for each object in a scene a data structure with the same neighborhood structure as the domain required for the task at hand. A change in point of view, for example, can then be easily accomplished diagrammatically by "rotating" the labels of the orientation with respect to the intrinsic one.

- We introduce heuristics to control the propagation of constraints by using the hierarchical and functional decomposition of space to limit constraints to physically adjacent objects. A weighting of positional relations according to their informational content is used to avoid "information decay" in the network due to the propagation of weak relations.

- We discuss a method for constraint relaxation that uses the structure of the relational domain to weaken constraints by including other neighboring relations in their disjunctive definitions, instead of retracting them as a whole. This approach leads faster to solutions of meaningfully modified sets of otherwise unsatisfiable constraints.

- We propose a comprehensive knowledge representation model, characterized by making the role of the observer explicit. We use the representation model to look into the various modalities of representation such as declarative, procedural, propositional, analogical, etc., and conclude, that in general, only some aspects of a representation correspond to a particular modality, depending on the level of abstraction considered. The concept of qualitativeness is found to be orthogonal to the modalities discussed.

9.2   Future research issues

Integrating metric knowledge

In order to emphasize the relevance of qualitative knowledge, we have been insisting on the "disadvantages" of quantitative representations. If quantitative information is available, however, it might be desirable to integrate it with qualitative information, because they might constrain each other allowing additional inferences to be made.

Simmons (1986) describes a system called *Quantity Lattice*, which exemplifies such an integration for a broad class of "commonsense" arithmetic inferences. The idea is to represent expressions as nodes (called quantities) in a directed graph, with arcs representing (ordinal) relationships among expressions. Upper and lower numeric bounds in form of real valued intervals are associated with each quantity. In the case of positional information, the arcs would represent qualitative topological and orientation relations, whereas the numerical intervals would represent ranges of compatible coordinates for the position of the objects represented as a node. The *Quantity Lattice* system is able to determine the relationship between two quantities and to constrain the value of the arithmetic expressions according to the information available using various reasoning techniques including graph search, numeric constraint propagation, and interval-, relational- and constant-elimination-arithmetic. Further investigations on the combination of quantitative and qualitative knowledge can be found in Williams (1988), Meiri (1991), MacNish, Galton, and Gooday (1993). The constraint reasoning mechanisms introduced in section 5.4 correspond to the graph search techniques mentioned here (enriched, however, by the structural constraints of the spatial domain), and could be extended in a similar manner to include numeric constraint propagation and arithmetic operations. This is certainly an important avenue of future research.

Integrating path knowledge

As a field of study, spatial cognition is often subdivided in the study of small scale and large scale environments (the criterion for small being if it can be perceived as a whole), and further in highly structured environments (such as a city) and unstructured environments with visible landmarks (such as a landscape). Depending on the task at hand (sketching a map, giving directions, active wayfinding), different sources of knowledge and diverse strategies are used, a fact reflected in multiple levels of representation. Even if we restrict ourselves to a simple two-dimensional world, there is more than one way of "knowing" where an object is. One is the static, object-centered perspective we have been studying. Another one is a dynamic, viewer-centered perspective based on distinctive landmarks and paths connecting them. Paths can be as simple as a sequence of straight segments connected by changes in direction, or be characterized by following an irregular route or using a means of transportation.

The qualitative approach relates in two ways to path knowledge. First, changes in direction in route descriptions are typically given in qualitative terms. Second, the positional information implicitly contained in paths can be gradually assimilated in form of topological and orientation relations by a process in which positional relations are established for the (virtual) places at which directional changes occur. The positions of the actual objects in the scene are then derived by constraint propagation.

Integrating non-spatial knowledge

As we saw in the section on visualization, a lot of non-spatial knowledge can provide useful additional constraints in the spatial reasoning process. Thus, a major concern for practical applications is how to integrate other sources of knowledge. An object-centered approach seems the most appropriate for the following reasons:

- Object-centered spatial representations are more stable, and thus better suited for long term memory, because they are independent of the viewer's position.

- Inheritance hierarchies allow the efficient storage of common properties by associating them with classes or prototype objects. Instances inherit those properties, but can supersede them if they happen to be special cases.

- The slot-filler associational structure of object-centered representations allow the flexible incorporation of new knowledge.

The disadvantages, such as that some types of knowledge might be distributed among many objects and be difficult to access, and that artificial pseudo-classes might seem to be necessary, can be usually overcome with additional indexing techniques and sound knowledge engineering.

Interpretation of relations and its relationship to linguistic concepts

Even though our approach is inspired in part by the way we describe space in natural language, the topological and orientation relations used model positional information and not linguistic locative prepositions. For example, many of the locative prepositions in natural language do not have unique inverses, which are required for the constraint propagation mechanism. Also the transitivity of spatial prepositions is restricted by many global and local constraints, for example, *left of* is locally, but not globally transitive (Habel and Pribbenow 1989). On the other hand, the topological and orientation relations are mathematical abstractions with well defined properties, among them the existence of inverse relations and bounded transitivity.[1]

With other words, at present, the approach is intended to be a model of our knowledge of space and *not* of how language describes space. However, both are tightly related and the computational framework can be adapted to model linguistic restrictions. For example, the fact that the composition tables have an entry for a given pair of relations does not imply that the composition will propagate without bounds. The mechanisms that use the tables to do constraint propagation can control how "far" a given relation can be propagated to avoid a general information decay due to the dissemination of weak information, and thus provide a similar restraining effect as in the linguistic case. Also, typical qualitative descriptions are hierarchically organized around distinguished reference objects, and it is up to the propagation algorithm to decide if and how far to propagate information across hierarchical boundaries.

Furthermore, as was said in chapter 6, the qualitative model could be used as the core of a system that accepts natural language input (and produces natural language output). The general procedure would be to translate the linguistic expressions into a canonical internal representation, do whatever processing is necessary, and then—through context-dependent language generation—translate back to the linguistic expressions (see Figure 6.1). This is the only way to cope with the context dependance of language, whereby the qualitative approach offers the possibility of maintaining the vagueness of language in the representation instead of forcing us to use ranges of values or the like.

---

[1]Transitivity is the special case of the composition of a relation with itself resulting in the same relation, e.g., A[b]B, B[b]C → A[b]C, and is a property of the relation.

Perceptual representations and mental models

In a similar vein to the previous point, it should be noted, that even though we have been freely using results from research originally intended to explain perceptual representations and mental models, in the context of developing artificial representations, these two are separate scientific aims. In particular, care should be taken not to assume that the representations we propose actually model the way spatial knowledge is represented in the mind.

Keeping that in mind, there is, of course, useful cross-fertilization among these two research fields that should be further explored. The study of perceptual representations can provide the "requirements" for cognitively adequate representations like, for example:

- Qualitativeness

- Hierarchical organization

- Ability to reason at different granularity levels (fine vs. coarse reasoning)

- Bounded short time memory

On the other hand, artificial knowledge representation approaches can provide *computational* models to operationally test the soundness of the theories about perceptual representations.

9.3 Summary

Most common approaches to the representation of spatial knowledge rely on Euclidean geometry, one of the most ancient axiomatizations of space, in one form or the other. That is, they assume points and the distance between points as the most basic notions from which other geometric sorts such as vectors, angles, regions, and volumes can be derived. The metric used to define distance can be a discrete one, or a continuous one. There are several reasons why such approaches are not adequate for applications that require models of space similar to those used by humans.

Humans have trouble determining and remembering exact quantities such as lengths and angles. Furthermore, the precision of exact quantities is not required, because cognition is "situated", that is, spatial representations are built and used in concrete situations. Also, evidence from psychological studies (e.g., Roberts and Suppes 1967) suggest that cognitive space is not Euclidian.

We argue in favor of the use of qualitative representations of commonsense knowledge in general, and spatial knowledge in particular. Qualitative representations make only those few distinctions of borderline cases, that are needed to discriminate among alternatives in a given context. Qualitative representations are particularly well suited for "situated" applications, where the immediate interaction with a "real world" makes them usefully constrained.

In order to represent the relative position of two objects in 2-D space qualitatively, we define a small set of spatial relations from the two relevant dimensions topology and orientation. The relative position is given by a topologi-
cal/orientation relation pair. Topological relations are derived from the combinatorial variations of the point set intersection of boundaries and interiors of
the involved objects by imposing the constraints of physical space on them. The
orientation dimension results from the transfer of distinguished reference axes
from an observer to the reference object. Relative orientations must be given
w.r.t. a reference frame, which can be intrinsic (orientation given by some inherent property of the reference object), extrinsic (orientation imposed by external
factors), or deictic (orientation imposed by point of view). When reasoning
about orientations, the reference frame is implicitly assumed to be the intrinsic
orientation of the parent object (i.e., the one containing the objects involved),
unless explicitly stated otherwise.

We present a variety of mechanisms to reason with qualitative representations in general, and qualitative representations of 2-D positional information in
particular. One of the simplest is transforming between explicit reference frames
(intrinsic, extrinsic, deictic) and a canonical implicit one. Another is computing the composition of spatial relations. While the composition of topological
and orientation relations result in relation sets that tend to contain too many
alternative relations, the composition of positional information, i.e., of topologi-
cal/orientation pairs, yields more specific results. Interestingly, only a fraction of
all possible variations of relations actually occur as compositions, and all of them
are connected. Also, "fixed-points" are reached after a few iterations. Simple
rules for the computation of composition can be given (for example, "shortest
path" rule for orientations). We survey the solution techniques for the general
constraint satisfaction problem available in the literature, and find them to be
"limited by their generality". We show that taking the structure of the richly
constrained spatial domain into consideration leads to more efficient algorithms
either by direct use of analogical data structures, or by improving the control of
the constraint propagation and constraint relaxation procedures.

Qualitative representations of space can be used in many application areas,
particularly in those characterized by uncertain and incomplete knowledge. We
discuss two applications in the realm of visual processing, demonstrating the
"knowledge assimilation" and "knowledge use" aspects of our representation,
respectively: building a "cognitive map" from single views of a scene in a simplified setting, and visualizing qualitatively described office layouts.

Finally, we point at ways to extend the qualitative approach to represent
positional information in 3-D scenes, as well as other spatial concepts such as
size, shape and distance.

Appendix A

Composition tables for various special cases

This appendix contains additional composition tables illustrating the regularities discussed in chapter 5. It concentrates on the composition of binary topological relations, as well as the combined composition of topological and orientation relations for solids, because the composition of orientation relations (for points) follows a simple rule (shortest path). The composition of orientations at different granularity levels follows the same rule provided the orientations are represented using internal ranges (see section 4.4.2). These tables were put together by Daniel Kobler, who also wrote the TeX-macros necessary to draw the icons used in them.

A∏B	B∏C							
	d	t	o	i⊕b	i	c⊕b	c	=
d	{d,t,o,i⊕b,i,c⊕b,c,=}	{d,t,o,i⊕b,i}	{d,t,o,i⊕b,i}	{d,t,o,i⊕b,i}	{d,t,o,i⊕b,i}	{d}	{d}	{d}
t	{d,t,o,c⊕b,c}	{d,t,o,i⊕b,c⊕b,=}	{d,t,o,i⊕b,i}	{t,o,i⊕b,i}	{o,i⊕b,i}	{d,t}	{d}	{t}
o	{d,t,o,c⊕b,c}	{d,t,o,c⊕b,c}	{d,t,o,i⊕b,i,c⊕b,c,=}	{o,i⊕b,i}	{o,i⊕b,i}	{d,t,o,c⊕b,c}	{d,t,o,c⊕b,c}	{o}
c⊕b	{d,t,o,c⊕b,c}	{t,o,c⊕b,c}	{o,c⊕b,c}	{o,i⊕b,c⊕b,=}	{o,i⊕b,i}	{c⊕b,c}	{c}	{c⊕b}
c	{d,t,o,c⊕b,c}	{o,c⊕b,c}	{o,c⊕b,c}	{o,c⊕b,c}	{o,i⊕b,i,c⊕b,c,=}	{c}	{c}	{c}
i⊕b	{d}	{d,t}	{d,t,o,i⊕b,i}	{i⊕b,i}	{i}	{d,t,o,i⊕b,c⊕b,=}	{d,t,o,i⊕b,c⊕b,c}	{i⊕b}
i	{d}	{d}	{d,t,o,i⊕b,i}	{i}	{i}	{d,t,o,i⊕b,i}	{d,t,o,i⊕b,i,c⊕b,c,=}	{i}
=	{d}	{t}	{o}	{i⊕b}	{i}	{c⊕b}	{c}	{=}

Table A.1: Composition of binary topological relations (set representation)

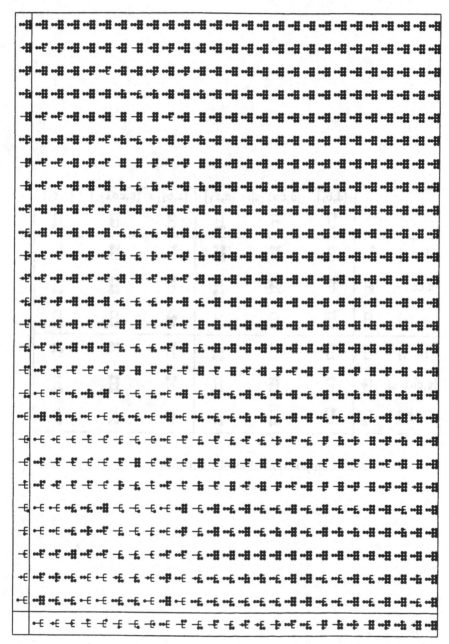

Table A.2: Full composition table for the closed set of binary topological relations

	$[d,b_1]$	$[d,\mathrm{col}_{fb}]$	$[d,f_1]$	$[t,b_1]$	$[t,\mathrm{col}_{fb}]$	$[t,f_1]$
$[d,b_1]$						
$[d,\mathrm{col}_{fb}]$						
$[d,f_1]$						
$[t,b_1]$						
$[t,\mathrm{col}_{fb}]$						
$[t,f_1]$						

Table A.3: Composition of level 1 orientations for solids

	$[\psi,\beta]$	[d,r]	[d,b]	[d,l]	[d,f]	[t,r]	[t,b]	[t,l]	[t,f]
$[\varphi,\alpha]$									
[d,r]									
[d,b]									
[d,l]									
[d,f]									
[t,r]									
[t,b]									
[t,l]									
[t,f]									

Table A.4: Composition of level 2 orientations for solids

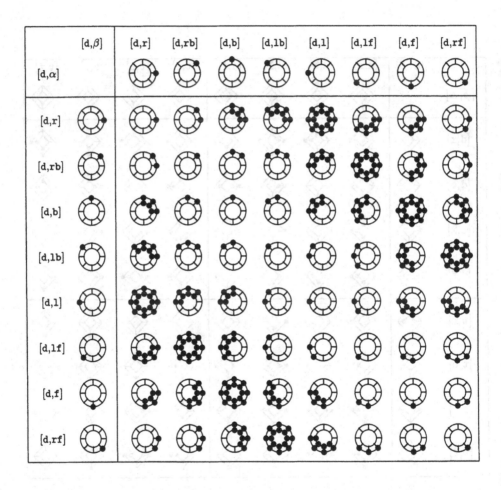

Table A.5: Composition of level 3 orientations for solids ([d,]/[d,] case)

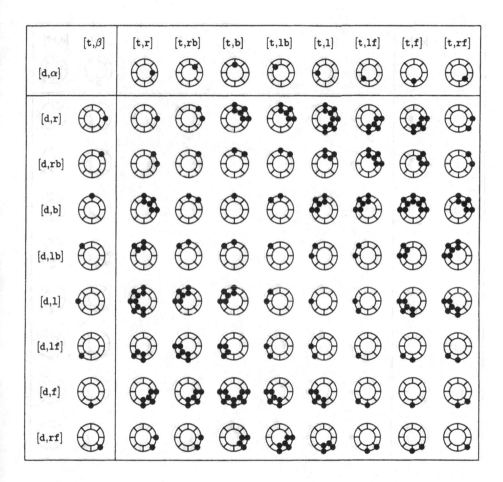

Table A.6: Composition of level 3 orientations for solids ([d,]/[t,] case)

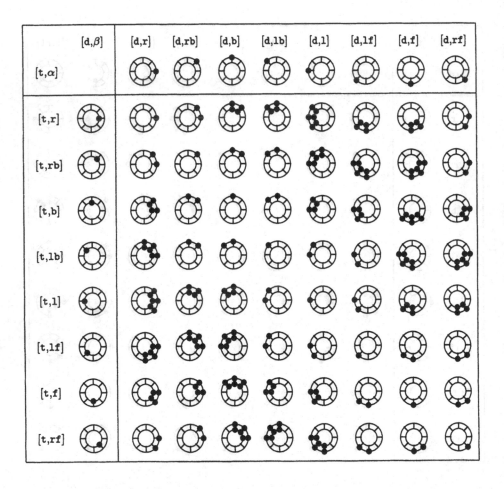

Table A.7: Composition of level 3 orientations for solids ([t,]/[d,] case)

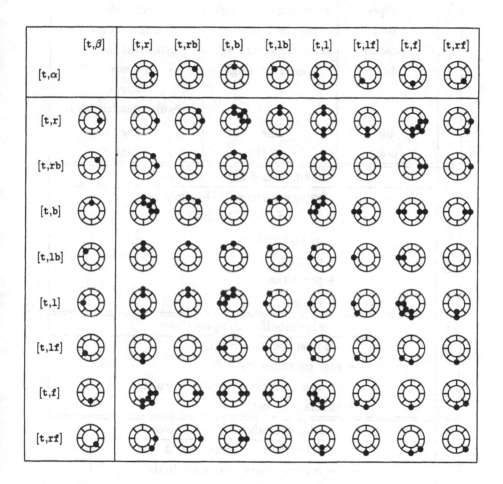

Table A.8: Composition of level 3 orientations for solids ([t,]/[t,] case)

	2D-primitives		
	round shapes	triangular shapes	quadrilateral shapes
l e v e l 1	circle ellipse oval scarab lens	isoceles and right-angled triangle isoceles triangle right-angled triangle irregular triangle	square rectangle parallelogram sym. trapezium rhomb kite irregular quadrilateral
	line point	line point	line point

l e v e l 2	not-modified 3D-primitives
	bottom-surface : 2D-primitive + top-surface : 2D-primitive + size-relation + position + projection + height

l e v e l 3	edge-modified 3D-primitives
	not-modified 3D-primitive + edge curvature (convex, concave, overall curvature) + auxiliary surface

l e v e l 4	surface-modified 3D-primitives
	3D-primitive of level 2 or 3 + surface - curvature - attribute (convex, concave)

l e v e l 5	measured 3D-primitives
	3D-primitive of level 2, 3 or 4 + qualitative measures (size, position, projection, compared, related)

Table A.9: Summary of shape description vocabulary

Bibliography

The numbers in brackets at the end of each entry indicate the pages in the text on which that entry is cited. This extended bibliography contains references to literature in the broader context of the subject matter, not all of which is explicitly cited in the text.

Abel, D. and Ooi, B. C., editors (1993). *Third International Symposium on Large Spatial Databases, SSD '93*, Volume 692 of *Lecture Notes in Computer Science*. Springer, Berlin. (169, 170, 175)

Abelson, H. and Sussman, G. J. (1985). *Structure and Interpretation of Computer Programs*. The MIT Press, Cambridge, MA. (122)

Adorni, G. and Di Manzo, M. (1983). Top down approaches to scene interpretation. In *Proceedings CIL 83*, Barcelona, Spain.

Adorni, G., Di Manzo, M., and Giunchiglia, F. (1983). Some basic mechanisms for common sense reasoning about stories environments. In 8th-IJCAI (1983), pages 72–74. (140)

Adorni, G., Di Manzo, M., and Giunchiglia, F. (1984). From descriptions to images: What reasoning in between? In O'Shea, T., editor, *Proceedings ECAI-84*, Pisa, Italy, pages 359–368.

Aiello, L. C. and Nardi, D. (1991). Research trends in knowledge representation. In Flach, P. A. and Meersman, R. A., editors, *Future Directions in Artificial Intelligence*, pages 83–92. North-Holland, Amsterdam. (143)

Allen, J. F. and Hayes, P. J. (1985). A common-sense theory of time. In Joshi (1985), pages 528–531. (130)

Allen, J. F. (1983). Maintaining knowledge about temporal intervals. *Communications of the ACM*, *26*(11), 832–843. (25, 78, 83, 85, 96, 119, 129, 130)

Ambler, A. P. and Popplestone, R. J. (1975). Inferring the positions of bodies from specified spatial relations. *Artificial Intelligence*, *6*, 129–156.

American Association for Artificial Intelligence, 5th-AAAI (1986). *Proceedings of the Fifth National Conference on Artificial Intelligence.* AAAI Press, Menlo Park. (171, 186, 189, 191)

American Association for Artificial Intelligence, 7th-AAAI (1988). *Proceedings of the Seventh National Conference on Artificial Intelligence.* AAAI Press, Menlo Park. (184, 191)

American Association for Artificial Intelligence, 8th-AAAI (1990). *Proceedings of the Eighth National Conference on Artificial Intelligence.* AAAI Press/The MIT Press, Menlo Park/Cambridge. (167, 168, 180, 181, 183, 184)

American Association for Artificial Intelligence, 9th-AAAI (1991). *Proceedings of the Ninth National Conference on Artificial Intelligence.* AAAI Press/The MIT Press, Menlo Park/Cambridge. (181, 183)

André, E., Bosch, G., Herzog, G., and Rist, T. (1987). Coping with the intrinsic and deictic uses of spatial prepositions. In Jorrand, P. and Sgurev, V., editors, *Artificial Intelligence II. Methodology, Systems, Applications*, pages 375–382. North-Holland, Amsterdam. (60)

André, E., Herzog, G., and Rist, T. (1988). On the simultaneous interpretation of real world image sequences and their natural language description: the system SOCCER. In Kodratoff, Y., editor, *Proc. of the 8th ECAI*, pages 449–454. Pitman, London.

Andress, K. M. and Kak, A. C. (1988). Evidence accumulation and flow of control in a hierarchical spatial reasoning system. *AI Magazine*, 9(2), 75–94.

Anger, F., Guesgen, H.-W., and van Benthem, J., editors (1993). *Proceedings of the Workshop on Spatial and Temporal Reasoning at the 13th International Joint Conference on Artificial Intelligence*, Chambéry, France. (166, 169, 181, 182, 184, 190)

Arkin, R. C. (1990). Integrating behavioral, perceptual, and world knowledge in reactive navigation. *Robotics and Autonomous Systems*, 6(1), 105–122.

Asher, N., Aurnague, M., Bras, M., and Vieu, L. (1993). Space, time and discourse. In Anger et al. (1993), pages 219–235.

Aurnague, M. and Vieu, L. (1993). Toward a formal representation of space in language: A commonsense reasoning approach. In Anger et al. (1993), pages 123–158.

Bäckström, C. (1990). Logical modeling of simplified geometrical objects and mechanical assembly processes. In Chen (1990), pages 35–61. (141)

Baird, J. (1979). Studies of the cognitive representation of spatial relations: I. overview. *Journal of Experimental Psychology: General*, 108(1), 90–91. (143)

Baird, J., Merrill, A., and Tannenbaum, J. (1979). Cognitive representation of spatial relations: A familiar environment. *Journal of Experimental Psychology: General*, *108*(1), 92–98. (143)

Bajcsy, R., editor (1993). *Proceedings of the Thirteenth International Joint Conference on Artificial Intelligence*, Chambéry, France. International Joint Conferences on Artificial Intelligence, Inc., Morgan Kaufmann, San Mateo, CA. (167, 174, 181, 188, 190)

Ballard, D. H. and Brown, C. M. (1982). *Computer Vision*. Prentice-Hall, Englewood Cliffs, NJ.

Barrow, H. G. and Popplestone, R. J. (1971). Relational descriptions in picture processing. *Machine Intelligence*, *6*, 377–396.

van Beek, P. (1990). Reasoning about qualitative temporal information. In 8th-AAAI (1990), pages 728–734.

Bennett, D. C. (1975). *Spatial and Temporal Uses of English Prepositions. An Essay in Stratificational Semantics*. Longman, London. (145)

Bergler, R. (1990). Allens Zeitlogik: Eine Scheme Implementation. Fortgeschrittenenpraktikum, Institut für Informatik, Technische Universität München. (83)

Beringer, A., Hölldobler, S., and Kurfeß, F. (1993). Spatial reasoning and connectionist inference. In Bajcsy (1993), pages 1352–1357.

Bestougeff, H. and Ligozat, G. F. (1992). Logical tools for temporal knowledge representation. In Campbell, J., editor, *Ellis Horwood in Artificial Intelligence*. Ellis Horwood, New York.

Bialystok, E. and Olson, D. R. (1987). Spatial categories: The perception and conceptualization of spatial relations. In Harnad, S., editor, *Categorial Perception: The Groundwork of Cognition*. Cambridge University Press, Cambridge, MA. (143)

Bierwisch, M. (1988). On the grammar of local prepositions. In Bierwisch, Motsch, and Zimmermann, editors, *Syntax, Semantik und Lexikon*, Number 29 in Studia Grammatica, pages 1–65. Akademie-Verlag, Berlin. (145)

Binford, T. O. (1971). Visual perception by computer. In *Proceedings IEEE Conf. Syst. Contr.*, Miami, Fl. (122)

Blades, M. (1991a). The development of the abilities required to understand spatial representations. In Mark and Frank (1991), pages 81–115. (143)

Blades, M. (1991b). The development of wayfinding abilities. In Mark and Frank (1991), pages 137–165. (143)

Blocher, A., Stopp, E., and Weis, T. (1992). ANTLIMA-1: Ein System zur Generierung von Bildvorstellungen ausgehend von Propositionen. SFB 314 (VITRA), Memo 50, Universität des Saarlandes, Saarbrücken, Germany.

Block, N., editor (1981). *Imagery*. MIT Press, Cambridge. (180, 186)

Bobrow, D., editor (1984). *Qualitative Reasoning about Physical Systems*. Elsevier Science Publishers, Amsterdam. (7, 129)

Bobrow, D. G. (1975). Dimensions of representation. In Bobrow, D. G. and Collins, A., editors, *Representation and Understanding*, chapter 1, pages 1–34. Academic Press, New York. (16)

Bock, J. K. (1982). Toward a cognitive psychology of syntax: Information processing contributions to sentence formulation. *Psychological Review, 89*, 1–47.

Borning, A. (1979). Thinglab: A constraint-oriented simulation laboratory. Report CS-79-746, Computer Science Dept., Stanford University, CA.

Brachman, R. J. and Levesque, H. J., editors (1985). *Readings in Knowledge Representation*. Morgan Kaufmann, San Mateo, CA. (176)

Brachman, R. J. and Smith, B. C. (1980). Special issue on knowledge representation. *Sigart newsletter, 70*, 41–137.

Brachman, R. J. (1990). The future of knowledge representation. In 8th-AAAI (1990), pages 1082–1092.

Bräunling, P. (1990). Zur kontextabhängigen Objektbenennung — Ist ein großes gelbes Haus immer groß und gelb? In Hoeppner (1990), pages 1–6. (119)

Bräunling, P., Freksa, C., and Zimmermann, K., editors (1990). The SpaceGarden Bibliography. Forschungsberichte Künstliche Intelligenz FKI-138-90, Institut für Informatik, Technische Universität München. (129)

Burger, W. and Bhanu, B. (1992). *Qualitative Motion Planning*. Kluwer, Dordrecht.

Carney, S. P. and Brown, D. C. (1989). A qualitative model for reasoning about shape and fit. In Gero, J. S., editor, *Artificial Intelligence in Design*, pages 251–273. Springer, Berlin.

Casati, R. and Varzi, A. C. (in press). *Holes and other Superficialities*. The MIT Press, Cambridge, MA. (122)

Chandrasekaran, B. and Narayanan, N. H. (1990). Towards a theory of commonsense visual reasoning. In Nori, K. V. and Madhavan, C. E. V., editors, *Foundations of Software Technology and Theoretical Computer Science, Tenth Conference*, Bengalore, India, Lecture Notes in Computer Science, pages 388–409. Springer, Berlin. (82, 140, 143)

Chandrasekaran, B. and Narayanan, N. H. (1992). Reasoning with diagrammatic representations. Proceedings of the 1992 AAAI Spring Symposium, March 25-27th, 1992, Stanford University. (21)

Chang, S.-K. and Jungert, E. (1990). A spatial knowledge structure for visual information systems. In Ichikawa, T., Jungert, E., and Korfhage, R., editors, *Visual Languages and Applications*, pages 277–304. Plenum Press, New York.

Chang, S.-K. and Li, Y. (1988). Representation of multi-resolution symbolic and binary pictures using 2DH-strings. In *Proceedings of the IEEE Workshop on Language for Automation*, pages 190–195. (147)

Chang, S.-K., Shi, Q.-Y., and Yan, C.-W. (1987). Iconic indexing by 2D-strings. *IEEE Transactions on Pattern Analysis and Machine Intelligence*, 9(3), 413–427. (147)

Charniak, E., Riesbeck, C. K., and McDermott, D. V. (1980). *Artificial Intelligence Programming*. Lawrence Erlbaum, Hillsdale, NJ. (96)

Chatila, R. (1982). Path planning and environment learning in a mobile robot system. In *Proceedings ECAI-82*, Paris.

Chen, L. (1982). Topological structure in visual perception. *Science*, 218(699).

Chen, L. (1989). Topological perception: A challenge to computational approaches to vision. In Pfeifer, R. et al., editors, *Connectionism in Perspective*, pages 317–329. North-Holland, Amsterdam.

Chen, S.-s., editor (1990). *Advances in Spatial Reasoning*, Volume 1. Ablex, Norwood, NJ. (166)

Clark, H. H. and Chase, W. G. (1972). On the process of comparing sentences against pictures. *Cognitive Psychology*, 3, 472–517. (134)

Clarke, B. (1981). A calculus of individuals based on "connection". *Notre Dame Journal of Formal Logic*, 22(3), 204–218. (139)

Clarke, B. (1985). Individuals and points. *Notre Dame Journal of Formal Logic*, 26(1), 61–75. (139)

Clementini, E., Di Felice, P., and van Oosterom, P. (1993). A small set of formal topological relationships suitable for end-user interaction. In Abel and Ooi (1993), pages 277–295. (138)

Cohen, R. (1983). *The Development of Spatial Cognition*. Lawrence Erlbaum, Hillsdale, NJ. (143)

Cohn, A. G. (1993). Modal and non-modal qualitative spatial logics. In Anger et al. (1993), pages 95–100.

Cohn, A. G., Cui, Z., and Randell, D. A. (1992). Logical and computational aspects of spatial reasoning. In Pribbenow, S. and Schlieder, C., editors, *Spatial Concepts: Connecting Cognitive Theories with Formal Representations*. ECAI-92 Workshop. Reprinted as: Bericht Graduiertenkolleg Kognitionswissenschaft, Universität Hamburg, 1993. (137)

Cohn, A. G., Randell, D. A., Cui, Z., and Bennett, B. (1993). Qualitative spatial reasoning and representation. In Piera Carreté and Singh (1993), pages 513–522.

Cowen, D., editor (1992). *Fifth International Symposium on Spatial Data Handling*, Charleston, NC. (172, 182, 186)

Cresswell, M. J. (1978). Prepositions and points of view. *Linguistics and Philosophy*, *2*(1), 1–41.

Cui, Z., Cohn, A. G., and Randell, D. A. (1992). Qualitative simulation based on a logical formalism of space and time. In *Proceedings of the Tenth National Conference on Artificial Intelligence*, pages 679–684. American Association for Artificial Intelligence, AAAI Press/The MIT Press, Menlo Park/Cambridge. (137)

Cui, Z., Cohn, A. G., and Randell, D. A. (1993). Qualitative and topological relationships in spatial databases. In Abel and Ooi (1993), pages 293–315.

Davis, E. (1983). The MERCATOR representation of spatial knowledge. In 8th-IJCAI (1983), pages 295–301. (144)

Davis, E. (1986). *Representing and Acquiring Geographic Knowledge*. Morgan Kaufmann, San Mateo, CA. (144)

Davis, E. (1987). A framework for qualitative reasoning about solid objects. In Rodriguez, G., editor, *Proceedings of the Workshop on Space Telerobotics*, pages 369–375. NASA and JPL. Reprinted in (Weld and de Kleer 1990b).

Davis, E. (1989). Solutions to a paradox of perceptual acuity. In *Proceedings of the First Conference on Theoretical Aspects of Knowledge Representation*, Toronto.

Davis, E. (1990). *Representations of commonsense knowledge*. Morgan Kaufmann, San Mateo, CA.

Davis, R., Shrobe, H., and Szolovits, P. (1993). What is a knowledge representation. *AI Magazine*, *14*(1), 17–33.

Dean, T. and Boddy, M. (1988). Reasoning about partially ordered events. *Artificial Intelligence*, *36*, 375–387. Reprinted in (Weld and de Kleer 1990b).

Dechter, A. and Dechter, R. (1987). Removing redundancies in constraint networks. In *Proceedings of the Sixth National Conference on Artificial Intelligence*, pages 105–109. American Association for Artificial Intelligence, AAAI Press, Menlo Park. (73)

Dechter, R. (1986). Learning while searching in constraint satisfaction problems. In 5th-AAAI (1986), pages 178–183. (74)

Dechter, R. (1990). Enhancement schemes for constraint processing: Backjumping, learning, and cutset decomposition. *Artificial Intelligence*, *41*(3), 273–312. (74)

Dechter, R. and Pearl, J. (1985). The anatomy of easy problems: A constraint satisfaction formulation. In Joshi (1985), pages 1066–1072.

Dechter, R. and Pearl, J. (1988). Network-based heuristics for constraint-satisfaction problems. *Artificial Intelligence*, *34*(1), 1–38. (73)

Di Manzo, M., Giunchiglia, F., and Pino, E. (1985). Space representation and object positioning in natural language driven image generation. In Bibel, W. and Petkoff, B., editors, *Artificial Intelligence Methodology, Systems, Applications*, Varna, Bulgaria, pages 207–214. North-Holland, Amsterdam. (135)

Dickinson, S. J. (1991). The recovery and recognition of three-dimensional objects using part-based aspect matching. Technical report CAR-TR-572, Center for Automation Research, University of Maryland, College Park, MD. (122, 123)

Downs, R. M. and Stea, D., editors (1973). *Image and Environment: Cognitive Mapping and Spatial Behaviour*. Aldine, Chicago. (176)

Doyle, J. (1979). A truth maintenance system. *Artificial Intelligence*, *12*, 231–272. (92, 96)

Edelsbrunner, H. (1987). *Algorithms in Combinatorial Geometry*. Springer, Berlin.

Eder, A. and Schneider, K. (1991). Anwendung des qualitativen Repräsentationsansatzes im Bereich des architektonischen Entwurfs. Fortgeschrittenenpraktikum, Institut für Informatik, Technische Universität München.

Edwards, G. (1991). Propositional and spatial knowledge for image understanding. In Mark and Frank (1991), pages 295–307.

Egenhofer, M. J. (1989). A formal definition of binary topological relationships. In Litwin, W. and Schek, H.-J., editors, *Third International Conference on Foundations of Data Organization and Algorithms*, Volume 367 of *Lecture Notes in Computer Science*, pages 457–472. Springer, Berlin. (130)

Egenhofer, M. J. (1991). Reasoning about binary topological relations. In Gunther, O. and Schek, H.-J., editors, *Advances in Spatial Databases, Second Symposium on Large Spatial Databases*, Volume 525 of *Lecture Notes in Computer Science*, pages 143–160. Springer, Berlin. (34, 61, 62, 63, 64, 130)

Egenhofer, M. J. and Al-Taha, K. K. (1992). Reasoning about gradual changes of topological relationships. In Frank et al. (1992), pages 196–219. (34, 130)

Egenhofer, M. J. and Franzosa, R. (1991). Point-set topological spatial relations. *International Journal of Geographical Information Systems*, 5(2), 161–174. (34, 35, 36)

Egenhofer, M. J. and Herring, J. (1990). A mathematical framework for the definition of topological relationships. In *Fourth International Symposium on Spatial Data Handling*, Zurich, Switzerland, pages 803–813.

Egenhofer, M. J. and Herring, J. (1991). High-level spatial data structures in GIS. In Maguire et al. (1991), pages 227–237.

Egenhofer, M. J. and Sharma, J. (1992). Topological consistency. In Cowen (1992), pages 335–343. (34)

Egenhofer, M. J. and Sharma, J. (1993). Assessing the consistency of complete and incomplete topological information. *Geographical Systems*, 1, 47–68. (34, 130)

Eldracher, M., Hernández, D., and Kinder, M. (1992). Concept of an integrated trajectory generation system. Forschungsberichte Künstliche Intelligenz FKI-171-92, Institut für Informatik, Technische Universität München.

Evans, T. G. (1968). A heuristic program to solve geometric analogy problems. In Minsky, M., editor, *Semantic Information Processing*. The MIT Press, Cambridge, MA. (47)

Faltings, B. (1990). Qualitative kinematics in mechanisms. *Artificial Intelligence*, 44(1), 89–119.

Feldman, J. A. (1985). Four frames suffice: A provisional model of vision and space. *Behavioral and Brain Sciences*, 8, 265–289.

Feldman, J. A. (1988). Time, space and form in vision. Technical report TR 244, University of Rochester, Computer Science Department.

Forbus, K. D. (1983). Qualitative reasoning about space and motion. In Gentner, D. and Stevens, A. L., editors, *Mental Models*, pages 53–73. Lawrence Erlbaum, Hillsdale, NJ.

Forbus, K. D. (1984). Qualitative process theory. *Artificial Intelligence*, 24, 85–168. (11, 17)

Forbus, K. D. (1988). Qualitative physics: Past, present, and future. In Shrobe, H., editor, *Exploring Artificial Intelligence*, pages 239–296. Morgan Kaufmann, San Mateo, CA. Reprinted in (Weld and de Kleer 1990b).

Forbus, K. D., Nielsen, P., and Faltings, B. (1991). Qualitative spatial reasoning: The clock project. *Artificial Intelligence*, *51*, 417–471. (136)

Frank, A. U. (1991). Qualitative spatial reasoning with cardinal directions. In Kaindl, H., editor, *7. Österreichische Artificial Intelligence Tagung*. Springer, Berlin.

Frank, A. U. (1992). Qualitative spatial reasoning with cardinal directions. *Journal of Visual Languages and Computing*, *3*, 343–371. (138)

Frank, A. U. and Campari, I., editors (1993). *Spatial Information Theory. A Theoretical Basis for GIS. European Conference, COSIT'93*, Marciana Marina, Italy, Volume 716 of *Lecture Notes in Computer Science*. Springer, Berlin. (177, 178, 179, 181, 182, 185, 191, 192)

Frank, A. U., Campari, I., and Formentini, U., editors (1992). *Theories and Methods of Spatio-Temporal Reasoning in Geographic Space. Intl. Conf. GIS—From Space to Territory*, Pisa, Volume 639 of *Lecture Notes in Computer Science*. Springer, Berlin. (172, 173, 179, 185)

Frank, A. U. and Mark, D. M. (1991). Language issues for geographical information systems. In Maguire et al. (1991), pages 147–163.

Franklin, N., Tversky, B., and Coon, V. (1992). Switching points of view in spatial mental models acquired from text. *Memory and Cognition*, *20*, 507–518. (60)

Freeman, J. (1975). The modelling of spatial relations. *Computer Graphics and Image Processing*, *4*, 156–171. (134)

Freksa, C. (1991). Qualitative spatial reasoning. In Mark and Frank (1991), pages 361–372. (8, 9, 136, 137)

Freksa, C. (1992a). Temporal reasoning based on semi-intervals. *Artificial Intelligence*, *54*, 199–227. (11, 64, 69, 130, 131, 132)

Freksa, C. (1992b). Using orientation information for qualitative spatial reasoning. In Frank et al. (1992), pages 162–178. (138)

Freksa, C. and Habel, C., editors (1990a). *Repräsentation und Verarbeitung räumlichen Wissens*. Informatik Fachberichte 245. Springer, Berlin. (173, 176, 186, 188)

Freksa, C. and Habel, C. (1990b). Warum interessiert sich die Kognitionsforschung für die Darstellung räumlichen Wissens? In Freksa and Habel (1990a). (1)

Freksa, C. and Hernández, D. (1991). Qualitative and quantitative knowledge about physical space. Unpublished position paper at the SPQR Workshop on Multiple Models, Karlsruhe.

Freksa, C. and Röhrig, R. (1993). Dimensions of qualitative spatial reasoning. In Piera Carreté and Singh (1993), pages 483–492. (138, 139)

Freksa, C. and Zimmermann, K. (1992). On the utilization of spatial structures for cognitively plausible and efficient reasoning. In *Proceedings of the 1992 IEEE International Conference on Systems, Man, and Cybernetics*, Chicago. (138, 139)

Freuder, E. C. (1978). Synthesizing constraint expressions. *Communications of the ACM*, *21*(11), 958–966. (72)

Freuder, E. C. (1982). A sufficient condition for backtrack-free search. *Journal of the Association for Computing Machinery*, *29*(1), 24–32. (73)

Fuhr, T., Kummert, F., Posch, S., and Sagerer, G. (1992). An approach for qualitatively predicting relations from relations. In Sandewall, E. and Jansson, C. G., editors, *Proceedings of the Scandinavian Conference on Artificial Intelligence*, pages 38–49. IOS Press, Amsterdam. (138)

Fujihara, H. and Mukerjee, A. (1991). Qualitative reasoning about document structures. Technical report TAMU 91-010, Computer Science Department, Texas A & M University. (40, 105, 133)

Funt, B. V. (1976). *WHISPER: A Computer Implementation Using Analogues in Reasoning*. Ph.D. thesis, University of British Columbia. Reprinted as Technical Report of the School of Computing Science, Simon Fraser University, Burnaby, B.C., V5A 1S6, Canada, 1992.

Funt, B. V. (1980). Problem solving with diagrammatic representations. *Artificial Intelligence*, *13*(3), 201–230. (22, 111, 118, 140, 143)

Funt, B. V. (1987). Analogical modes of reasoning and process modelling. In Cercone, N. and McCalla, G., editors, *The Knowledge Frontier—Essays in the Representation of Knowledge*, chapter 15, pages 414–428. Springer, Berlin.

Furbach, U., Dirlich, G., and Freksa, C. (1985). Towards a theory of knowledge representation systems. In Bibel, W. and Petkoff, B., editors, *Artificial Intelligence Methodology, Systems, Applications*. North-Holland, Amsterdam. (17)

Galton, A. (1993). Towards an integrated logic of space, time, and motion. In Bajcsy (1993), pages 1550–1555.

Gapp, K.-P. (1993). Berechnungsverfahren für räumliche Relationen in 3D-Szenen. Diplomarbeit, Universität des Saarlandes, Saarbrücken, Germany. Published as: SFB 314 (VITRA), Memo Nr. 59.

Gardin, F. and Meltzer, B. (1989). Analogical representations of naive physics. *Artificial Intelligence*, *38*, 139–159. (143)

Glasgow, J. I. and Papadias, D. (1992). Computational imagery. *Cognitive Science*, *16*, 355–394.

Gotts, N. M. (1987). A qualitative spatial representation for cardiac electro-physiology. In Fox, J., Fieschi, M., and Engelbrecht, R., editors, *Lecture Notes in Medical Informatics*, Volume 33, pages 88–95. Springer, Berlin.

Gould, M. D. (1991). Integrating spatial cognition and geographic information systems. In Mark and Frank (1991), pages 435–447.

Green, S. (1987). SPACES — A system for the representation of commonsense knowledge about space for design. In Bramer, M. A., editor, *Research and Development in Expert Systems III*, pages 195–206. Cambridge University Press, Cambridge, MA. (136)

Guesgen, H.-W. (1989). Spatial reasoning based on Allen's temporal logic. Technical report TR-89-049, ICSI, Berkeley, CA. (25, 130, 132)

Güting, R. (1988). Geo-relational algebra: A model and query language for geometric database systems. In Schmidt, J. W. et al., editors, *Advances in Database Technology EDBT 88, International Conference on Extending Database Technology*, pages 506–527. Springer, Berlin. (34, 146)

Güting, R. and Schneider, M. (1993). Realms: A foundation for spatial data types in database systems. In Abel and Ooi (1993), pages 14–35.

Haar, R. (1976). Computational models of spatial relations. Technical report TR-478 MCS-72-0361-10, Dept. of Computer Science, University of Maryland. (47)

Habel, C., Herweg, M., and Rehkämper, K., editors (1989). *Raumkonzepte in Verstehensprozessen*. Niemeyer, Tübingen. (178, 186)

Habel, C. (1988a). Cognitive linguistics: The processing of spatial concepts. LILOG-Report 45, IBM Deutschland. (145)

Habel, C. (1988b). Prozedurale Aspekte der Wegplanung und Wegbeschreibung. In Schnelle, H. and Rickheit, G., editors, *Sprache in Mensch und Computer*, pages pp.107–133. Westdeutscher Verlag, Wiesbaden. (145)

Habel, C. (1988c). Repräsentation räumlichen Wissens. In Rahmstorf, G., editor, *Wissensrepräsentation in Expertensystemen*. Springer, Berlin. (145)

Habel, C. (1989). Propositional and depictorial representations of spatial knowledge: The case of path concepts. Mitteilung 171, Univ. Hamburg. Fachbereich Informatik. (140, 145)

Habel, C. (1990). Repräsentation von Wissen. *Informatik Spektrum*, *13*(3), 126–136.

Habel, C. and Pribbenow, S. (1988). Gebietskonstituierende Prozesse. LILOG-Report 18, IBM Deutschland. (32)

Habel, C. and Pribbenow, S. (1989). Zum Verstehen räumlicher Ausdrücke des Deutschen — Transitivität räumlicher Relationen. In Brauer, W. and Freksa, C., editors, *Wissensbasierte Systeme, 3. Internationaler GI-Kongreß*, München, pages 139–152. Springer, Berlin. (143, 145, 152)

Hafner, W. and Kobler, D. (1991). Repräsentation räumlichen Wissens: stabile Raumdarstellung. Fortgeschrittenenpraktikum, Institut für Informatik, Technische Universität München. (107, 109, 110, 112)

Hahn, T. (1990). Funts parallele Retina zur Handhabung diagrammatischer Repräsentationen. Fortgeschrittenenpraktikum, Institut für Informatik, Technische Universität München. (111)

Haller, S. M. and Mark, D. M. (1990). Knowledge representation for understanding geographical locatives. In *Fourth International Symposium on Spatial Data Handling*, Zurich, Switzerland, pages 465–477.

Hart, R. A. and Moore, G. T. (1973). The development of spatial cognition: A review. In Downs and Stea (1973). (111, 143)

Haugeland, J. (1987). An overview of the frame problem. In Pylyshyn (1987), pages 77–94. (140)

Havens, W. and Mackworth, A. (1983). Representing knowledge of the visual world. *IEEE Computer*, *16*(10), 90–96.

Hayes, P. J. (1974). Some problems and non-problems in representation theory. In *Proceedings AISB Summer Conference*, pages 63–79. University of Sussex. Reprinted in (Brachman and Levesque 1985).

Hayes, P. J. (1979). The naive physics manifesto. In Michie, D., editor, *Expert Systems in the Electronic Age*, pages 242–270. Edinburgh University Press, Edinburgh, Scotland. (11)

Hayes, P. J. and Allen, J. F. (1987). Short time periods. In *Proceedings of the Tenth International Joint Conference on Artificial Intelligence*, Milan, Italy. International Joint Conferences on Artificial Intelligence, Inc., Morgan Kaufmann, San Mateo, CA.

Hays, E. (1990). On defining motion verbs and spatial prepositions. In Freksa and Habel (1990a), pages 192–206. (145)

Hernández, D. (1984). Modulare Softwarebausteine zur Wissensrepräsentation. Studienarbeit, IMMD (IV) and RRZE Universität Erlangen-Nürnberg. (96)

Hernández, D. (1989). Zur Implementierbarkeit analogischer Repräsentationen. In Metzing (1989), pages 479–481.

Hernández, D. (1991). Relative representation of spatial knowledge: The 2-D case. In Mark and Frank (1991), pages 373–385.

Hernández, D. (1992). Diagrammatical aspects of qualitative representations of space. In Narayanan, N. H., editor, *Proceedings of the 1992 AAAI Spring Symposium on Reasoning with Diagrammatic Representations, March 25-27th, 1992*, Stanford University, CA, pages 225–228.

Hernández, D. (1993a). Hybride und integrierte Ansätze zur Raumrepräsentation und ihre Anwendung. In Herzog, O., Christaller, T., and Schütt, D., editors, *Grundlagen und Anwendungen der Künstlichen Intelligenz. 17. Fachtagung für Künstliche Intelligenz*, Humboldt-Universität zu Berlin, Informatik Aktuell, pages 210–216. Springer, Berlin.

Hernández, D. (1993b). Maintaining qualitative spatial knowledge. In Frank and Campari (1993), pages 36–53.

Hernández, D., editor (1993c). Proceedings des Workshops "Hybride und integrierte Ansätze zur Raumrepräsentation und ihre Anwendung". Forschungsberichte Künstliche Intelligenz FKI-185-93, Institut für Informatik, Technische Universität München. Im Rahmen der 17. Fachtagung für Künstliche Intelligenz vom 13.-16. Sept. 1993 an der Humboldt-Universität zu Berlin.

Hernández, D. (1993d). Reasoning with qualitative representations: Exploiting the structure of space. In Piera Carreté and Singh (1993), pages 493–502.

Hernández, D. and Zimmermann, K. (1993). Default reasoning and the qualitative representation of spatial knowledge. Forschungsberichte Künstliche Intelligenz FKI-175-93, Institut für Informatik, Technische Universität München. To appear in Habel et al. *Defaults and Prototypes—Non-monotonic Reasoning for Language and Knowledge Processing*, Springer. (100)

Herring, J. (1991). The mathematical modeling of spatial and non-spatial information in geographic information systems. In Mark and Frank (1991), pages 313–350.

Herskovits, A. (1986). *Language and Spatial Cognition. An Interdisciplinary Study of the Prepositions in English*. Cambridge University Press, Cambridge, MA. (145)

Hertzberg, J., Güsgen, H.-W., Voß, A., Fidelak, M., and Voß, H. (1988). Relaxing constraint networks to resolve inconsistencies. In Hoeppner, W., editor, *GWAI-88, 12. Jahrestagung Künstliche Intelligenz. Proceedings*, Eringerfeld, Informatik - Fachberichte 181. Springer, Berlin. (100)

Herweg, M. (1989). Ansätze zu einer semantischen Beschreibung topologischer Präpositionen. In Habel et al. (1989). (145)

Hill, C. (1982). Up/down, front/back, left/right. A contrastive study of Hausa and English. In Weissenborn, J. and Klein, W., editors, *Here and There. Cross-Linguistic Studies on Deixis and Demonstration*. John Benjamins, Amsterdam. (145)

Hinrichs, E. W. (1986). A compositional semantics for directional modifiers— locative case reopened. In *Proceedings of COLING*, Bonn.

Hirtle, S. C. (1991). Knowledge representations of spatial relations. In Doignon, J.-P. and Falmagne, J.-C., editors, *Mathematical Psychology: Current developments*, pages 233–250. Springer, Berlin.

Hirtle, S. C., Ghiselli-Crippa, T., and Spring, M. S. (1993). The cognitive structure of space: An analysis of temporal sequences. In Frank and Campari (1993), pages 177–189.

Hoeppner, W., editor (1990). *Workshop Räumliche Alltagsumgebungen des Menschen, Bericht 9/90*. Universität Koblenz-Landau. (168, 188)

Högg, S. (1993). Ein Modell zur qualitativen Beschreibung von Grundformen. Diplomarbeit, Institut für Informatik, Technische Universität München. (122, 124, 125, 126)

Högg, S. and Schwarzer, I. (1991). Composition of spatial relations. Forschungsberichte Künstliche Intelligenz FKI-163-91, Institut für Informatik, Technische Universität München. (44)

Hunter, J. R. W., Gotts, N. M., Hamlet, I., and Kirby, I. (1989). Qualitative spatial and temporal reasoning in cardiac electrophysiology. In *AIME-89: Proceedings of the Second European Conference on Artificial Intelligence in Medicine*. Springer, Berlin.

International Joint Conferences on Artificial Intelligence, Inc., 5th-IJCAI (1977). *Proceedings of the Fifth International Joint Conference on Artificial Intelligence*, Cambridge, USA. Morgan Kaufmann, San Mateo, CA. (182, 190)

International Joint Conferences on Artificial Intelligence, Inc., 8th-IJCAI (1983). *Proceedings of the Eighth International Joint Conference on Artificial Intelligence*, Karlsruhe, FRG. Morgan Kaufmann, San Mateo, CA. (165, 170, 182)

Joe, G. and Mukerjee, A. (1990). Qualitative spatial representation based on tangency and alignments. Technical report TAMU 90-015, Computer Science Department, Texas A & M University.

Johnson-Laird, P. N. (1983). *Mental Models*. Harvard University Press, Cambridge, MA.

Johnson-Laird, P. N. (1988). *The Computer and the Mind.* Harvard University Press, Cambridge, MA. (9)

Joshi, A., editor (1985). *Proceedings of the Ninth International Joint Conference on Artificial Intelligence*, Los Angeles, CA. International Joint Conferences on Artificial Intelligence, Inc., Morgan Kaufmann, San Mateo, CA. (165, 171)

Jungert, E. (1988). Extended symbolic projections as a knowledge structure for spatial reasoning. In Kittler, J., editor, *4th International Conference on Pattern Recognition*, Volume 301 of *Lecture Notes in Computer Science*, pages 343–351. Springer, Berlin. (147)

Jungert, E. (1992). The observer's point of view: An extension of symbolic projections. In Frank et al. (1992), pages 179–195. (147)

Jungert, E. (1993a). Rotation invariance in symbolic projections as a means for determination of binary object relations. In Piera Carreté and Singh (1993), pages 503–512. (147)

Jungert, E. (1993b). Symbolic spatial reasoning on object shapes for qualitative matching. In Frank and Campari (1993), pages 444–462. (147)

Jungert, E. and Chang, S.-K. (1989). An algebra for symbolic image manipulation and transformation. In Kunii, T., editor, *Visual Database Systems*, pages 301–317. North-Holland, Amsterdam.

Kainz, W. (1990). Spatial relationships—topology versus order. In *Fourth International Symposium on Spatial Data Handling*, Zurich, Switzerland, pages 814–819.

Kainz, W., Egenhofer, M., and Greasly, I. (in press). Modeling spatial relations and operations with partially ordered sets. *International Journal of Geographical Information Systems*, 7.

Kautz, H. A. (1985). Formalizing spatial concepts and spatial language. In Hobbs, J. R. et al., editors, *Commonsense Summer: Final Report. No. CSLI-85-35.* Center for the Study of Language and Information, Stanford University.

Khenkhar, M. (1989). DEPIC-2D: Eine Komponente zur depiktionalen Repräsentation und Verarbeitung räumlichen Wissens. In Metzing (1989), pages 318–322. (140)

Kim, S. E. (1988). Viewpoint: Toward a computer for visual thinkers. Technical report STAN-CS-88-1190, Dept. of Computer Science, Stanford University, CA.

King, J. S. and Mukerjee, A. (1990). Inexact visualization: Qualitative object representation for recognizable reconstruction. In *Proceedings of the IEEE Conference on Biomedical Visualization*, Atlanta, GA, pages 136–143. (133)

de Kleer, J. (1986). An assumption based truth maintenance system. *Artificial Intelligence*, *28*, 127–162. (92)

Klein, W. (1983). Deixis and spatial orientation in route directions. In Pick and Acredolo (1983).

Kobler, D. (1991). Die Generierung einer stabilen Raumdarstellung. Forschungsberichte Künstliche Intelligenz FKI-161-91, Institut für Informatik, Technische Universität München. (108, 112)

Kobler, D. (1992). Visualisierung qualitativer Repräsentationen räumlichen Wissens. Diplomarbeit, Institut für Informatik, Technische Universität München. (44, 49, 115)

Koczy, L. T. (1988). On the description of relative position of fuzzy patterns. *Pattern Recognition Letters*, *8*, 21–28.

Kosslyn, S., Pinker, S., Smith, G., and Shwartz, S. (1981). On the demystification of mental imagery. In Block (1981), pages 131–150. (140)

Kosslyn, S. M. (1990). Mental imagery. In Osherson et al. (1990), chapter 3, pages 73–97.

Kramer, G. A. (1990). Solving geometric constraint systems. In 8th-AAAI (1990), pages 708–714. (70)

Kuipers, B. J. (1978). Modelling spatial knowledge. *Cognitive Science*, *2*, 129–153. (144)

Kuipers, B. J. and Byun, Y. T. (1987). A qualitative approach to robot exploration and map learning. In *Proceedings of the IEEE Workshop on Spatial Reasoning and Multi-Sensor Fusion*, pages 390–404. Morgan Kaufmann, San Mateo, CA. (144)

Kuipers, B. J. and Levitt, T. (1988). Navigation and mapping in large-scale space. *AI Magazine*, *9*(2), 25–43. (32, 143)

Kumar, V. (1992). Algorithms for constraint satisfaction problems: A survey. *AI Magazine*, *13*(1), 32–44. (70)

Lakoff, G. (1987). *Women, Fire, and Dangerous Things: What Categories Reveal About the Mind.* University of Chicago Press, Chicago.

Lakoff, G. and Johnson, M. (1980). *Metaphors we live by.* University of Chicago Press, Chicago.

Lang, E. (1987). Semantik der Dimensionsauszeichnung räumlicher Objekte. In Bierwisch, M. and Lang, E., editors, *Grammatische und konzeptuelle Aspekte von Dimensionsadjektiven*. Akademie-Verlag, Berlin. (145)

Lang, E. and Carstensen, K.-U. (1989). OSKAR: Ein PROLOG-Programm zur Modellierung der Struktur und der Verarbeitung räumlichen Wissens. In Metzing (1989), pages pp.234–243. (134)

Latecki, L. and Röhrig, R. (1993). Orientation and qualitative angle for spatial reasoning. In Bajcsy (1993), pages 1544–1549.

Laurini, R. and Milleret, F. (1988). Spatial data base queries: Relational algebra versus computational geometry. In Rafamelli et al., editors, *Proceedings of the Fourth International Conference on Statistical and Scientific Database Management*, Rome, Italy, pages 291–313. Springer, Berlin. (146)

Laurini, R. and Thompson, D. (1992). *Fundamentals of Spatial Information Systems*. Number 37 in The A.P.I.C. Series. Academic Press, New York.

Lee, S.-Y. and Hsu, F.-J. (1991). Picture algebra for spatial reasoning of iconic images represented in 2D C-string. *Pattern Recognition Letters*, *12*, 425–435. (147)

Levelt, W. J. M. (1986). Zur sprachlichen Abbildung des Raumes: Deiktische und intrinsische Perspektive. In Bosshardt, H.-G., editor, *Perspektiven auf Sprache*. de Gruyter, Berlin. (60, 145)

Levitt, T. and Lawton, D. (1990). Qualitative navigation for mobile robots. *Artificial Intelligence*, *44*, 305–360. (32, 144)

Lienhardt, P. (1991). Topological models for boundary representations: a comparison with n-dimensional generalized maps. *Computer Aided Design*, *23*(1), 59–82.

Ligozat, G. F. (1990). Weak representations of interval algebras. In 8th-AAAI (1990), pages 715–720. (130)

Ligozat, G. F. (1991). On generalized interval calculi. In 9th-AAAI (1991), pages 234–240.

Ligozat, G. F. (1993a). Models for qualitative spatial reasoning. In Anger et al. (1993), pages 35–45. (139)

Ligozat, G. F. (1993b). Qualitative triangulation for spatial reasoning. In Frank and Campari (1993), pages 54–68. (139)

Ligozat, G. F. and Bestougeff, H. (1989). On relations between intervals. *Information Processing Letters*, *32*, 177–182.

Lindsay, R. K. (1988). Images and inference. *Cognition*, *29*, 229–250. (140)

Lynch, K. (1960). *The Image of the City*. The MIT Press, Cambridge, MA.

Maaß, W. (1993). A cognitive model for the process of multimodal, incremental route descriptions. In Frank and Campari (1993), pages 1–13.

Maaß, W., Wazinski, P., and Herzog, G. (1993). Vitra guide: Multimodal route descriptions for computer assisted vehicle navigation. In *Sixth International Conference on Industrial & Engineering Applications of Artificial Intelligence & Expert Systems*, Edinburgh, pages 104–112.

Mackworth, A. K. (1977a). Consistency in networks of relations. *Artificial Intelligence*, 8, 99–118.

Mackworth, A. K. (1977b). On reading sketch maps. In 5th-IJCAI (1977). (22)

Mackworth, A. K. (1987). Constraint satisfaction. In Shapiro (1987). (70)

Mackworth, A. K. (1988). Adequacy criteria for visual knowledge representation. In Pylyshyn, Z. W., editor, *Computational Processes in Human Vision*, pages 462–474. Ablex Publishing Co., Norwood, NJ.

MacNish, C., Galton, A., and Gooday, J. (1993). Combining qualitative and quantitative reasoning. In Anger et al. (1993), pages 101–105. (151)

Maguire, D. J., Goodchild, M. F., and Rhind, D. W., editors (1991). *Geographical Information Systems: Principles and Applications*. Longman Scientific and Technical, London. (172, 173)

Malik, J. and Binford, T. O. (1983). Reasoning in time and space. In 8th-IJCAI (1983), pages 343–345.

Mani, K. and Johnson-Laird, P. N. (1982). The mental representation of spatial descriptions. *Memory and Cognition*, 10(2). (143)

Mark, D. M. (1992). Spatial metaphors for human-computer interaction. In Cowen (1992), pages 104–112.

Mark, D. M. and Egenhofer, M. J. (1992). An evaluation of the 9-intersection for region-line relations. In *GIS/LIS '92*, San Jose, CA, pages 513–521. (61)

Mark, D. M. et al. (1989). Working bibliography on "languages of spatial relations". Report 89-10, National Center for Geographic Information and Analysis, Santa Barbara, CA. (129)

Mark, D. M. and Frank, A. U. (1989). Concepts of space and spatial language. In *Proceedings Ninth International Symposium on Computer-Assisted Cartography (Auto-Carto 9)*, Baltimore, MD, pages 538–556.

Mark, D. M. and Frank, A. U., editors (1991). *Cognitive and Linguistic Aspects of Geographic Space*. NATO Advanced Studies Institute. Kluwer, Dordrecht. (167, 171, 173, 175, 177)

Mark, D. M., Svorou, S., and Zubin, D. (1987). Spatial terms and spatial concepts: Geographic, cognitive, and linguistic perspectives. In *Proceedings International Symposium on Geographic Information Systems: The Research Agenda*, Crystal City, VA, Volume 2, pages 101–112.

Marr, D. (1982). *Vision*. Freeman, New York. (16)

Maturana, H. R. and Varela, F. J. (1980). *Autopoiesis and Cognition*. Reidel, Dordrecht. (13)

Maturana, H. R. and Varela, F. J. (1984). *El Árbol del Conocimiento. Las bases biológicas del entendimiento humano*. Editorial Universitaria, Santiago de Chile.

Mavrovouniotis, M. L. and Stephanopoulus, G. (1988). Formal order-of-magnitude reasoning in process engineering. *Computer Chemical Engineering*, *12*, 867–880. (11, 120)

McAllester, D. (1980). An outlook on truth maintenance. Memo 551, MIT Artificial Intelligence Laboratory, Cambridge, MA. (92)

McAllester, D. (1990). Truth maintenance. In 8th-AAAI (1990), pages 1109–1116.

McCarthy, J. and Hayes, P. (1969). Some philosophical problems from the standpoint of artificial intelligence. In Meltzer, B. and Michie, D., editors, *Machine Intelligence 4*, pages 463–502. American Elsevier, New York.

McDermott, D. V. (1983). Contexts and data dependencies: A synthesis. *IEEE Transactions on Pattern Analysis and Machine Intelligence*, *5*(3), 237–246. (92)

McDermott, D. V. (1987). Reasoning, spatial. In Shapiro (1987), pages 863–870.

McDermott, D. V. (1992). Reasoning, spatial. In Shapiro, E., editor, *Encyclopedia of Artificial Intelligence* (Second edition)., pages 1322–1334. Wiley. (129)

McDermott, D. V. and Davis, E. (1984). Planning routes through uncertain territory. *Artificial Intelligence*, *22*, 107–156. (144)

McNamara, T. P. (1986). Mental representations of spatial relations. *Cognitive Psychology*, *18*(1), 87–121. (143)

Meiri, I. (1991). Combining qualitative and quantitative constraints in temporal reasoning. In 9th-AAAI (1991), pages 260–267. (151)

Meseguer, P. (1989). Constraint satisfaction problems: An overview. *AI Communications*, *2*(1), 3–17. (70)

Metzing, D., editor (1989). *Proceedings GWAI 1989*, Volume 216 of *Informatik Fachberichte*. Springer, Berlin. (177, 179, 181)

Minsky, M. (1986). *The Society of Mind*. Simon and Schuster, New York. (117)

Mitra, D. and Loganantharaj, R. (1993). Experimenting with a temporal constraint propagation algorithm. In Anger et al. (1993), pages 245–256.

Mohnhaupt, M. (1987). On modelling events with an analogical representation. In Morik, K., editor, *Proceedings GWAI 1987 11th German Workshop on Artificial Intelligence, Geseke*, pages 31–41. Springer, Berlin. (140)

Mohnhaupt, M. (1992). *Prinzipien piktorieller Repräsentationssysteme*, Volume 300 of *Informatik Fachberichte*. Springer, Berlin.

Montanari, U. (1974). Network of constraints: Fundamental properties and applications to picture processing. *Information Science*, *7*, 95–132. (72)

Mukerjee, A. (1991a). Accidental alignments: an approach to qualitative vision. In *IEEE Conference on Robotics and Automation*, Sacramento, CA, pages 1096–1101. (133)

Mukerjee, A. (1991b). Qualitative geometric design. In *Solid Modeling Foundations and CAD/CAM Applications, ACM/SIGGRAPH Symposium*, Austin, TX. (134)

Mukerjee, A. and Bratton, S. E. (1991). Vocabulary choice and modeling bias: A study in spatial learning. Technical report TAMU 91-009, Computer Science Department, Texas A & M University. (133)

Mukerjee, A. and Joe, G. (1990). A qualitative model for space. In 8th-AAAI (1990), pages 721–727. (119, 130, 133)

Neumann, B. and Mohnhaupt, M. (1988). Propositionale und analoge Repräsentation von Bewegungsverläufen. *KI*, *1*, 4–10.

Nielsen, P. (1988). A qualitative approach to mechanical constraint. In 7th-AAAI (1988), pages 270–274. Reprinted in (Weld and de Kleer 1990b).

Nilsson, N. J. (1969). A mobile automaton: An application of AI techniques. In *Proceedings of the First International Joint Conference on Artificial Intelligence*, Washington, D.C. International Joint Conferences on Artificial Intelligence, Inc., Morgan Kaufmann, San Mateo, CA.

Nökel, K. (1989). Convex relations between time intervals. In Retti, J. and Leidlmair, K., editors, *5. Österreichische Artificial-Intelligence-Tagung*, Volume 208 of *Informatik Fachberichte*, pages 298–302. GI, Springer, Berlin. (11, 129)

Nudel, B. (1983). Consistent-labeling problems and their algorithms: Expected complexities and theory-based heuristics. *Artificial Intelligence*, *21*, 135–178. (70, 74)

Nudel, B. (1988). Tree search and arc consistency in constraint satisfaction algorithms. In Kanal, L. N. and Kumar, V., editors, *Search in Artificial Intelligence*, pages 287–342. Springer, Berlin. (74)

Nutter, J. T. (1987). Reasoning, default. In Shapiro (1987). (9)

Olson, D. R. and Bialystok, E. (1983). *Spatial Cognition—The Structure and Development of Mental Representations of Spatial Relations*. Lawrence Erlbaum, Hillsdale, NJ. (143)

Osherson, D. N., Kosslyn, S. M., and Hollerbach, J. M., editors (1990). *An Invitation to Cognitive Science: Visual Cognition and Action*, Volume 2. The MIT Press, Cambridge, MA. (180)

Paivio, A. (1983). The empirical case for dual coding. In Yuille, J., editor, *Imagery, Memory, and Cognition*, pages 307–332. Lawrence Erlbaum, Hillsdale, NJ. (80)

Palmer, S. E. (1975a). The effects of contextual scenes on the identification of objects. *Memory and Cognition*, *3*(5), 519–526.

Palmer, S. E. (1975b). The nature of perceptual representation: An examination of the analog, propositional controversy. In Schank, R. and Nash-Webber, B. L., editors, *TINLAP-1*, pages 165–173. (20)

Palmer, S. E. (1978). Fundamental aspects of cognitive representation. In Rosch, E. and Lloyd, B. B., editors, *Cognition and Categorization*. Lawrence Erlbaum, Hillsdale, NJ. (13, 16, 21)

Papadias, D. and Sellis, T. (1992). Spatial reasoning using symbolic arrays. In Frank et al. (1992), pages 153–161. (147)

Papadias, D. and Sellis, T. (1993). The semantics of relations in 2D space using representative points: Spatial indexes. In Frank and Campari (1993), pages 234–247. (147)

Pederson, E. (1993). Geographic and manipulable space in two tamil linguistic systems. In Frank and Campari (1993), pages 294–311.

Pentland, A. P. (1986). Perceptual organization and the representation of natural form. *Artificial Intelligence*, *28*, 293–331. (122)

Peuquet, D. (1986). The use of spatial relationships to aid spatial database retrieval. In *Second International Symposium on Spatial Data Handling*, Seattle, WA, pages 459–471.

Peuquet, D. and Ci-Xiang, Z. (1987). An algorithm to determine the directional relationship between arbitrarily shaped polygons in the plane. *Pattern Recognition*, *20*(1), 65–74. (47, 49)

Pfefferkorn, C. E. (1975). A heuristic problem solving design system for equipment or furniture layouts. *Communications of the ACM*, *18*(5), 286–297. (105, 135)

Pick, H. L. and Acredolo, L. P., editors (1983). *Spatial Orientation: Theory, Research and Application*. Plenum Press, New York. (180, 190)

Piera Carreté, N. and Singh, M. G., editors (1993). *Proceedings of the III IMACS International Workshop on Qualitative Reasoning and Decision Technologies—QUARDET'93—*, Barcelona. CIMNE, Barcelona. (170, 174, 177, 179, 188)

Pigot, S. (1992). A topological model for a 3D spatial information system. In Cowen (1992).

Posner, M. I. (1989). *Foundations of Cognitive Science*. The MIT Press, Cambridge, MA.

Preparata, F. P. and Shamos, M. I. (1985). *Computational Geometry: an Introduction*. Springer, Berlin.

Pribbenow, S. (1989). Regelbasierte Interpretation lokaler Präpositionen am Beispiel von IN und BEI. In Habel et al. (1989).

Pribbenow, S. (1990). Interaktion von propositionalen und bildhaften Repräsentationen. In Freksa and Habel (1990a). (82, 140, 145)

Pribbenow, S. (1991). *Zur Verarbeitung von Lokalisierungsausdrücken in einem hybriden System*. Ph.D. thesis, Universität Hamburg. Reprinted as IWBS Report 211, IBM Deutschland, 1992. (60, 145)

Pylyshyn, Z. (1981). The imagery debate: Analog media vs tacit knowledge. In Block (1981), pages 151–206. (140)

Pylyshyn, Z., editor (1987). *The Robot's Dilemma: The Frame Problem in Artificial Intelligence*. Ablex, Norwood, NJ. (176)

Raiman, O. (1986). Order of magnitude reasoning. In 5th-AAAI (1986), pages 100–104. (120)

Randell, D. A. and Cohn, A. G. (1989). Modelling topological and metrical properties of physical processes. In Brachman, R., Levesque, H., and Reiter, R., editors, *Proceedings 1st International Conference on the Principles of Knowledge Representation and Reasoning*, pages 55–66. Morgan Kaufmann, San Mateo, CA.

Randell, D. A., Cohn, A. G., and Cui, Z. (1992a). Computing transitivity tables: A challenge for automated theorem provers. In *Proceedings CADE 11*. Springer, Berlin. (61)

Randell, D. A., Cohn, A. G., and Cui, Z. (1992b). An interval logic for space based on "connection". In Neumann, B., editor, *ECAI-92: Proceedings of the 10th European Conference on Artificial Intelligence*, pages 394–398. John Wiley, Chichester.

Randell, D. A., Cohn, A. G., and Cui, Z. (1992c). Naive topology: Modelling the force pump. In Faltings, B. and Struss, P., editors, *Recent Advances in Qualitative Physics*, pages 177–192. The MIT Press, Cambridge, MA. (137)

Randell, D. A., Cui, Z., and Cohn, A. G. (1992d). A spatial logic based on regions and connection. In *Proceedings 3rd International Conference on Knowledge Representation and Reasoning*, pages 165–176. Morgan Kaufmann, San Mateo, CA. (137)

Rehkämper, K. (1987). Mentale Bilder und Wegbedeutungen. LILOG-Report 20, IBM Deutschland, Stuttgart.

Rehkämper, K. (1988). Mentale Bilder — Analoge Repräsentationen. LILOG-Report 65, IBM Deutschland, Stuttgart.

Rehkämper, K. (1991). *Sind mentale Bilder bildhaft? — Eine Fragestellung zwischen Philosophie und Wissenschaft*. Ph.D. thesis, Universität Hamburg. (17)

Reiter, R. and Mackworth, A. K. (1989). A logical framework for depiction and image interpretation. *Artificial Intelligence, 41*, 125–155. (140, 141, 142)

Reitman, W. and Wilcox, B. (1988). Perception and representation of spatial relations in a program for playing go. In Levy, D. N. L., editor, *Computer Games*, Volume II, chapter 5.7, pages 192–202. Springer, Berlin. Bibliography at the end of book.

Retz-Schmidt, G. (1988). Various views on spatial prepositions. *AI Magazine, 9*(2), 95–105. (41, 44, 60, 145)

Rich, E. and Knight, K. (1991). *Artificial Intelligence* (2nd edition). McGraw-Hill, New York.

Roberts, F. S. and Suppes, P. (1967). Some problems in the geometry of visual perception. *Synthese, 17*, 173–201. (153)

Robinson, V. B. and Wong, R. N. (1987). Acquiring approximate representations of some spatial relations. In Chrisman, N. R., editor, *AUTO-CARTO 8, Eighth International Symposium on Computer-Assisted Cartography*, Baltimore, MD, pages 604–622.

Röhrig, R. (1993). CYCORD: a theory of qualitative spatial reasoning. Technical report LKI-M-93/6, Labor für Künstliche Intelligenz, Universität Hamburg, Hamburg, Germany. (139)

Rosenfeld, A. (1979). Digital topology. *American Mathematical Monthly*, *86*, 621–630.

Samet, H. (1990). *The Design and Analysis of Spatial Data Structures*. Addison-Wesley, Reading, MA.

Sander, R. (1991). Die Repräsentation räumlichen Wissens und die Behandlung von Einbettungsproblemen mit Quadtreedepiktionen. IWBS Report 191, IBM Deutschland. (78, 140)

de Saussure, F. (1916). *Cours de linguistique générale*. v.C. Bally and A. Sechehaye (eds.), Paris/Lausanne. English translation: Course in General Linguistics. London: Peter Owen, 1960. (18)

Schirra, J. R. J. (1991). Zum Nutzen antizipierter Bildvorstellungen bei der sprachlichen Szenenbeschreibung. SFB 314 (VITRA), Memo 49, Universität des Saarlandes, Saarbrücken, Germany.

Schirra, J. R. J. (1992). Connecting visual and verbal space: Preliminary considerations concerning the concept 'mental image'. In *4th European Workshop "Semantics of Time, Space and Movement and Spatio-Temporal Reasoning"*, Chateau de Bonas, France.

Schirra, J. R. J. and Stopp, E. (1993). ANTLIMA—A listener model with mental images. In Bajcsy (1993), pages 175–180.

Schlieder, C. (1990a). Anordnung. Eine Fallstudie zur Semantik bildhafter Repräsentation. In Freksa and Habel (1990a). (29, 31)

Schlieder, C. (1990b). Aquisition räumlichen Wissens am Beispiel ebener Sicht- und Anordnungsverhältnisse. In Hoeppner (1990), pages 159–171. (29, 32)

Schlieder, C. (1991). Anordnungswissen: Grundlage und Anwendung. Unpublished handout from a talk at the Tech. Univ. Munich. (29, 30)

Schlieder, C. (1993). Representing visible locations for qualitative navigation. In Piera Carreté and Singh (1993), pages 523–532.

Schwartz, D. G. (1989). Outline of a naive semantics for reasoning with qualitative linguistic information. In Sridharan, N. S., editor, *Proceedings of the Eleventh International Joint Conference on Artificial Intelligence*, Detroit, MI, Volume 2, pages 1068–1073. International Joint Conferences on Artificial Intelligence, Inc., Morgan Kaufmann, San Mateo, CA. (141)

Schwarzer, I. (1993). Qualitative Beschreibung zusammengesetzter Formen. Diplomarbeit, Institut für Informatik, Technische Universität München. (122, 127)

Schwarzer, I. and Högg, S. (1991). Komposition räumlicher Relationen. Fortgeschrittenenpraktikum, Institut für Informatik, Technische Universität München. (118)

Shapiro, E., editor (1987). Encyclopedia of Artificial Intelligence. Wiley. (182, 183, 185)

Simmons, R. (1986). "Commonsense" arithmetic reasoning. In 5th-AAAI (1986), pages 118–124. Reprinted in (Weld and de Kleer 1990b). (10, 150)

Sloman, A. (1971). Interactions between philosophy and AI — The role of intuition and non-logical reasoning in intelligence. In Proceedings of the Second International Joint Conference on Artificial Intelligence, London, UK, pages 270–278. International Joint Conferences on Artificial Intelligence, Inc., Morgan Kaufmann, San Mateo, CA. Reprinted in Artificial Intelligence (2) 1971. (20, 21, 140)

Sloman, A. (1975). Afterthoughts on analogical representation. In Proceedings Theoretical Issues in Natural Language Processing, Cambridge, MA, pages 164–168. (21, 22)

Sloman, A. (1978). Intuition and analogical reasoning. In Sloman, A., editor, The Computer Revolution in Philosophy: Philosophy, Science and Models of Mind, chapter 7. Harvester Press.

Sloman, A. (1984). Why we need many knowledge representation formalisms. In Bramer, M., editor, Research and Development in Expert Systems, Proceedings Expert Systems Conference. British Computer Society, Cambridge University Press, Cambridge, MA.

Smith, T. R. and Park, K. K. (1992). Algebraic approach to spatial reasoning. International Journal of Geographical Information Systems, 6(3), 177–192. (130)

Stallman, R. and Sussman, G. J. (1977). Forward reasoning and dependency-directed backtracking. Artificial Intelligence, 9(2), 135–196. (74, 92)

Steele, G. L. (1980). The Definition and Implementation of a Computer Programming Language Based on Constraints. Ph.D. thesis, MIT. Published as Report AI-TR-595. (73)

Steels, L. (1990). Exploiting analogical representations. Robotics and Autonomous Systems, 6(1), 71–88. (143)

Stevens, A. and Coupe, P. (1978). Distortions in judged spatial relations. Cognitive Psychology, 10, 422–437. (143)

Stiles-Davis, J., Kritchesvsky, M., and Bellugi, U., editors (1988). *Spatial Cognition - Brain Bases and Development.* Lawrence Erlbaum, Hillsdale, NJ. (143)

Stopp, E. (1993). GEO-ANTLIMA: Konstruktion dreidimensionaler mentaler Bilder aus sprachlichen Szenenbeschreibungen. Diplomarbeit, Universität des Saarlandes, Saarbrücken, Germany. Published as: SFB 314 (VITRA), Memo Nr. 60.

Struss, P. (1987). Problems of interval-based qualitative reasoning. Technical report INF 2 ARM-1-87, Siemens. A revised short version is reprinted in (Weld and de Kleer 1990b).

Sutherland, I. E. (1965). Sketchpath: A man-machine graphical communication system. Technical report 296, MIT Lincoln Lab, Cambridge, MA.

Talmy, L. (1983). How language structures space. In Pick and Acredolo (1983).

Thompson, A. M. (1977). The navigation system of the JPL robot. In 5th-IJCAI (1977).

Tolba, H., Charpillet, F., and Haton, J. P. (1991). Representing and propagating constraints in temporal reasoning. *AI Communications, 4*(4), 145–151.

Tolba, H. (1993). A contribution to the study of temporal reasoning: Integrating qualitative-quantitative information and propagation of constraints. In Anger et al. (1993), pages 283–296.

Topaloglou, T. (1991). Representation and management issues for large spatial knowledge bases. Technical report, Dept. of Computer Science, University of Toronto, Toronto, Ontario. (129, 146)

Touretzky, D. (1986). *The Mathematics of Inheritance Systems.* Morgan Kaufmann, San Mateo, CA.

Tversky, B. (1991). Spatial mental models. In Bower, G. H., editor, *The Psychology of Learning and Motivation: Advances in Research and Theory,* Volume 27, pages 109–145. Academic Press, New York. (143)

Ullman, S. (1985). Visual routines. In Pinker, S., editor, *Visual Cognition.* The MIT Press, Cambridge, MA.

Varzi, A. C. (1993). Spatial reasoning in a holey world: a sketch. In Anger et al. (1993), pages 47–59.

du Verdier, F. (1993). Solving geometric constraint satisfaction problems for spatial planning. In Bajcsy (1993), pages 1564–1569. (70)

Vieu, L. (1991). *Sémantique des relations spatiales et inférences spatio-temporelles.* Ph.D. thesis, Université Paul Sabatier, Toulouse.

Vieu, L. (1993). A logical framework for reasoning about space. In Frank and Campari (1993), pages 25–35. (139)

Vilain, M., Kautz, H., and van Beek, P. (1986). Constraint propagation algorithms for temporal reasoning: A revised report. In 5th-AAAI (1986), pages 377–382. Revised version reprinted in (Weld and de Kleer 1990b). (83, 84, 85)

Waltz, D. L. (1975). Understanding line drawings of scenes with shadows. In Winston (1975), pages 19–91. (66)

Waltz, D. L. and Boggess, L. (1979). Visual analog representations for natural language understanding. In *Proceedings of the Sixth International Joint Conference on Artificial Intelligence*, Tokio, Japan, pages 926–934. International Joint Conferences on Artificial Intelligence, Inc., Morgan Kaufmann, San Mateo, CA. (141)

Wazinski, P. (1993). Graduated topological relations. SFB 314 (VITRA), Memo 54, Universität des Saarlandes, Saarbrücken, Germany. (34)

Weld, D. S. and de Kleer, J. (1990a). Qualitative physics: A personal view. In Weld and de Kleer (1990b).

Weld, D. S. and de Kleer, J., editors (1990b). *Readings in Qualitative Reasoning about Physical Systems*. Morgan Kaufmann, San Mateo, CA. (7, 129, 170, 173, 184, 189, 190, 191)

Williams, B. C. (1986). Doing time: Putting qualitative reasoning on firmer ground. In 5th-AAAI (1986), pages 105–112. Revised version reprinted in (Weld and de Kleer 1990b).

Williams, B. C. (1988). MINIMA A symbolic approach to qualitative algebraic reasoning. In 7th-AAAI (1988), pages 264–269. Reprinted in (Weld and de Kleer 1990b). (151)

Winograd, T. and Flores, F. (1986). *Understanding Computers and Cognition*. Addison-Wesley, Reading, MA. (17)

Winston, P. H., editor (1975). *The Psychology of Computer Vision*. McGraw-Hill, New York. (134, 191)

Wunderlich, D. (1982). Sprache und Raum. *Studium Linguistik*, *12*, 1–19.

Wunderlich, D. and Herweg, M. (1989). Lokale und Direktionale. In von Stechow, A. and Wunderlich, D., editors, *Handbuch der Semantik*. Athenäum, Königstein.

Yeap, W. K., Taylor, P. S., and Jefferies, M. E. (1990). Computing a representation of the physical environment — A memory-based approach. AI Memo AIM-10-90-5, AI Lab University of Otago, New Zealand.

Yeap, W. K. (1988). Towards a computational theory of cognitive maps. *Artificial Intelligence*, *34*, 297–360. (107, 145)

Zadeh, L. A. (1965). Fuzzy sets. *Information Control*, *8*, 338–353. (134)

Zimmermann, K. (1991). SEqO: Ein System zur Erforschung qualitativer Objektrepräsentationen. Forschungsberichte Künstliche Intelligenz FKI-154-91, Institut für Informatik, Technische Universität München. (119)

Zimmermann, K. (1993). Enhancing qualitative spatial reasoning—combining orientation and distance. In Frank and Campari (1993), pages 69–76. (138)

Zimmermann, K. and Freksa, C. (1993). Enhancing spatial reasoning by the concept of motion. In Sloman, A., Hogg, D., Humphreys, G., Ramsay, A., and Partridge, D., editors, *Prospects for Artificial Intelligence, Proceedings of AISB93*, University of Birmingham. IOS Press, Amsterdam.

Index

Emphasized page numbers point to the definition or the main source of information about whatever is being indexed. An 'n' following the page number indicates that the indexed term appears in a footnote on that page.

Springer-Verlag
and the Environment

We at Springer-Verlag firmly believe that an international science publisher has a special obligation to the environment, and our corporate policies consistently reflect this conviction.

We also expect our business partners – paper mills, printers, packaging manufacturers, etc. – to commit themselves to using environmentally friendly materials and production processes.

The paper in this book is made from low- or no-chlorine pulp and is acid free, in conformance with international standards for paper permanency.

Printing: Weihert-Druck GmbH, Darmstadt
Binding: Theo Gansert Buchbinderei GmbH, Weinheim

Lecture Notes in Computer Science

Lecture Notes in Artificial Intelligence (LNAI)